Building Estimating

Building Estimating

George B. Wynne

Wake Technical Institute

CHARLES E. MERRILL PUBLISHING COMPANY

A Bell & Howell Company

Columbus, Ohio

Published by
Charles E. Merrill Publishing Co.
A Bell & Howell Company
Columbus, Ohio 43216

International Standard Book Number: 0–675–09037–7

Library of Congress Catalog Card Number: 72–88944

2 3 4 5 6 7–79 78 77 76 75
Printed in the United States of America

This Book is Dedicated in Loving
Memory to the Son of the Author,
CAPTAIN GEORGE TYLER WYNNE,
U.S. Field Artillery,
Who died February 14, 1970, in the 28th
Year of His Life While in the
Service of His Country.

It is Hoped That This Work
May Prove to be of Help to
Other Young Men Who are Seeking
Their Proper Destiny in the Complex
Society of This Wonderful Land of
Ours—a Land Worthy of Their Best
Efforts.

Foreword

From the beginning it should be understood that an estimate is just exactly what it says it is—only an estimate, which actually means a close approximation in which the experience of an estimator plays an important role in that it allows him to compare the particular estimate which he is making with estimates of like structures that he has estimated and seen built.

Because many more jobs are estimated than are actually built by any one contractor it follows that no more money should be spent on an estimate than is absolutely necessary; on the other hand, enough care should be exercised so that the price bid for each job would be low enough to be competitive and at the same time high enough to reasonably assure a profit or at the worst to prevent a loss.

The following article, which is reproduced here by permission of the editors of *Consulting Engineer*, points out some interesting facts about the art of estimating.

It is not intended that any of the rules set up in this article should be followed in making a detailed estimate. But this article supports the fact that an estimate, however skillfully made, is still an approximation.

Engineering Economics
The One-Dollar Estimate

Most things that weigh more than 100 pounds or that contain more than 100 cubic feet of something cost either $1.00/lb or $2.00/cu ft, whichever is more.
  This startling economic revelation can be applied to automobiles or engine-generator sets; to refrigerators or crawler tractors; to dishwashers or motor

graders; to sewage treatment plants or swimming pools. In short, it can be applied to just about everything that an engineer needs to estimate the cost of.

For example:

¶ A swimming pool, 20′ × 40′ × 5′ average depth, will cost about $8000, or $2.00/cu ft.

¶ A 2000 sq ft house in a suburban area will cost about $32,000, or $2.00/cu ft.

¶ A 4000-lb American-made automobile will carry a list price of about $4000.

¶ A 25,000-lb motor grader will sell for about $29,000, or $1.16/lb.

¶ A garden tractor weighing 630 lbs will cost about $730, also $1.16/lb.

The accompanying graph plots cost vs weight for a variety of different items, ranging from a 145-lb air compressor to a 54 cu yd scraper weighing

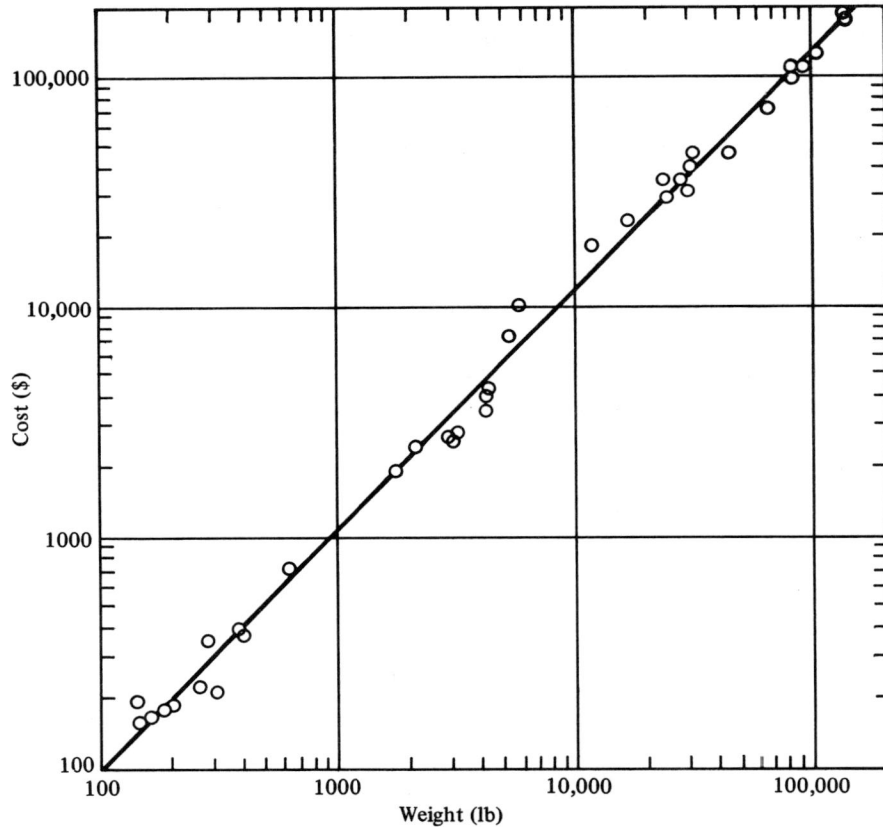

190,000 lbs. The list, covering 37 items, includes a dishwasher, washer and dryer, stove, refrigerator, power saw, furniture, automobiles, diesel engine-generator sets, scrapers, graders, and crawler tractors.

These items are, obviously, of quite utilitarian nature. As such, they constitute just one economic class of products. There are at least five different value levels that apply to materials or products in general: (1) aesthetic or luxury—more than $100/lb; (2) precision—$10/lb to $100/lb; (3) mechanical or electrical—$2/lb to $10/lb; (4) utilitarian or functional—$0.50/lb to $2.00/lb; and (5) bulk—less than $0.50/lb.

The fourth class—the $1.00/lb or $2.00/cu ft category—is of primary interest to the engineer in his predesign or order-of-magnitude estimates.

The last group—bulk items—includes materials and commodities that most individuals would be reluctant to accept delivery on, unless they had an immediate use in mind. A carload of nitrogenous sewage sludge, even though costing only $0.00125/lb on the open market, would be considered a distinct liability by most people.

The aesthetic or luxury class of product is usually available only in limited quantities, and the cost is largely independent of either weight or physical dimensions. Anything that costs more than $100/lb can be considered a luxury item. There are some interesting examples falling in this group.

The 69.42-carat diamond ring that recently sold at auction for $1.05 million, and was resold a few days later, works out to be priced in the neighborhood of $34.4 million/lb. That, however, is a neighborhood that few engineers need be concerned with.

Contact lenses offer a good example of a luxury or status item. At their common retail price of $200 per pair, contact lenses are going at $2.76 million/lb. Even at their wholesale price of $15.00/pair, they bring $207,000/lb, and their manufacturing cost of $1.00/pair still amounts to a substantial $13,800/lb.

Most engineering firms probably have a few items in the precision, perhaps even the luxury, class. A good theodolite lists at $116/lb, and a surveyor's transit will sell for $55/lb. Many electronic and optical instruments will cost from $10/lb to $100/lb.

The other cate cry—mechanical products—ranges in cost from $2.00/lb to $10.00/lb. In this group, a large part of the product's weight is made up of either electrical or mechanical parts. Many small appliances fit here.

Water, probably the cheapest readily-available commodity, usually costs less than $0.00005/lb; purified ribonuclease, the most expensive product, sells for $175 million/lb. But for most manufactured industrial products, the $1.00/lb estimate will be accurate within 30 percent.

William R. Park
Senior Engineering Economist
Midwest Research Institute

Mr. Park's article appeared on pages 114, 116 of *Consulting Engineer,* December, 1969.

Preface

This book has been prepared in an effort to present the principles of estimating in a clear and concise manner and to illustrate logical, systematic procedures in taking material from the plans, recording this material, and pricing the estimate. There has been a studied effort to stay away from the reference book idea of exhaustive discussions of materials and procedures and voluminous cost tables and charts.

All prices used in the examples in this text are intended to be 1971 national average prices unless otherwise stated. However, it must be borne in mind that material and equipment prices as well as labor rates will vary considerably from one section of the country to another. For this reason anyone preparing an estimate should ascertain the local prices for labor, materials, and equipment. But most important of all it must be remembered that the main purpose of this text is to teach methods and procedures and not exact prices.

The method employed in this text makes use of examples. Sketches are shown in the text from which the material is taken and then priced from information given in tables showing national average labor times and wages. Special attention is given to the detailed study of overhead and profit. After each material of construction is thus presented, a homework problem is assigned involving a similar item. A sketch is also provided in the text for the homework problem. One blank copy of a Quantity Takeoff, Summaries & Unit Costs, Direct Costs, and Overhead & Profit sheet is provided in Appendix F, pages 286-289. These forms can be duplicated if the instructor wishes.

The first item investigated is that of excavation. Then the principal structural materials are taken up: concrete, structural steel, masonry, and wood.

The last example shown in the text is for wood construction. For this example a set of house plans is shown. This example not only covers wood but also all of the other items required to make a complete house estimate such as excavation, roofing, finishing, mechanical, etc.

This text is intended to be flexible enough to cover one or two quarters of instruction. If a longer course is desired there is enough material developed in the latter chapters to prepare the student to estimate a building. A set of building plans was not included because it was felt the student should work with a set of plans relating to the particular area of the country in which he lives and will in all likelihood work.

June, 1973   George B. Wynne

Acknowledgements

The author gratefully acknowledges considerable assistance from Bethlehem Steel Company in the form of technical information.

¶ Mr. David Rogers, of Raleigh, North Carolina, an expert in the field of insurance and advanced financial planning, was very kind and generous with advice in all of the topics relating to his field of work.

¶ Mr. Willis Parker, Dean of Technical Education at Wake Technical Institute, Raleigh, North Carolina, was encouraging in many ways in the preparation of this work.

The following sources were very helpful in the furnishing of various types of information: *Engineering News Record;* Robert Snow Means Co., Inc.; Associated Equipment Distributors of Chicago; Frank R. Walker Company; North Carolina Concrete Masonry Association; Delmar Publishers, A Division of Litton Educational Publishing, Inc.; *Consulting Engineer;* National Lumber

Products Association; Associated General Contractors of America; Unit Structures Dept. of Koppers Company, Inc.; Cast A Stone Products Company; Carolina Builders Corporation; Pittsburgh Plate Glass Company; Peden Steel Company; the architectural firm of F. Carter Williams Architects; Robert B. Lyons, a teaching architect; and others.

Contents

CHAPTER THREE
Excavation

25

CHAPTER FOUR
Concrete

49

CHAPTER THIRTEEN
Special Equipment

CHAPTER FOURTEEN
Walks, Drives, and Parking Lots

CHAPTER FIFTEEN
Mechanical

APPENDICES

Building Estimating

CHAPTER ONE

The Mechanics of Estimating

In this text we will be primarily concerned with the proper practices and procedures necessary to prepare a reliable estimate. The cost of materials, labor, and equipment varies from one place to another and will have to be locally ascertained for each estimate. However, the times required to perform certain tasks do not vary very much from place to place so that tables for this item will be furnished.

ESTABLISHING COSTS

In practice these tables should be supplemented or replaced by the experience records of the estimator's company when such information is available. These experience records of the times required to accomplish a given task are important because every outfit varies in the efficiency its work force achieves even in comparison with other firms in the same area. There is a tendency, where pay is low, to give mediocre work response, while on the other hand higher pay usually encourages better performance habits but can foster arrogance and slowdowns. There is no valid substitute for actual experience records.

When the estimator has determined from the plans the proper amount of materials, labor, and equipment necessary for the project, and has applied the local prices to the quantities, he will have the basic cost to which must be added general office and job overhead expense and finally the anticipated profit in order to arrive at the bid price.

TYPES OF ESTIMATES

There are four distinct types of estimates, each of which has its place as the project moves from the promotional stage to acceptance and final payment.

However, it should be understood that this text will only concern itself with the detailed estimate. The four types are:

1. Detailed estimates
   a. Unit quantity method
   b. Total quantity method
2. Architect's or approximate method
3. Complete estimates
4. Periodic estimates
   a. Monthly
   b. Final

Detailed Estimates

The detailed estimate is a complete takeoff of all quantities of materials, labor, and equipment, and so on, and their pricing. This is the only reliable method and can be accomplished in two ways.

The Unit Quantity Method. The unit quantity method consists of finding the cost of a unit of each item and multiplying this figure by the number of units. This is done for each different item and the totals found. For example, the cost per cubic yard of concrete including all material, labor, and equipment is found and multiplied by the number of cubic yards of footing concrete required for the job; this is repeated for the concrete in the columns, in the beams for each floor, in the slabs, and so on. The cost per pound of structural steel for columns, beams, etc., is found and multiplied by the number of pounds in each of these required for the job. This is done for all items and the sum of the items contained in the entire project is computed. This is a good way to compare costs of the same items on different jobs and also to correct the estimate more easily for changes in quantities.

The Total Quantity Method. The total quantity method consists of finding, pricing, and bringing to a grand total all the materials, labor, and equipment that make up the job. For example, when there is concrete in the project the total quantities of each of the different materials that affect the cost of concrete are found. Such items include: the number of cubic yards of concrete, the pounds of reinforcing steel, the number of board feet of lumber for forms, and so on, along with the labor and equipment. These are added together to get the cost of the concrete. This is done for all of the materials in the job, which are then totaled. Overhead and profit are then added.

Architects' Estimates

The architect's estimate is employed by the architect in the planning stages of a project in order to determine whether or not the client can obtain the type of facility he desires for the amount of money that is available. This can save the architect from completing a set of plans that will bring in bids in excess of the desired cost. If the plans thus prove too expensive the architect is obliged to make extensive revisions in the plans at his own expense, which of course he wishes to avoid. This estimate is usually done by the square foot or cubic foot method or by a brief unit cost approach or a combination of the two. This type of estimate is useful in the early planning stages of proposed projects as a guide to approximate costs.

Complete Estimates

The complete estimate includes everything connected with the project in addition to the actual construction. It encompasses such items as: real estate, financing, insurance, lawyer's fees, taxes, and architect's fees, as well as the contract for the work.

Periodic Estimates

Estimates are made at different periods of the work.

Monthly Estimates. A monthly estimate is required to determine the amount of payment due the contractor for that part of the contract then completed. Ten percent or more of this amount is usually withheld to help insure satisfactory completion of the work.

Final Estimates. A final estimate is made at the completion of the job in order to determine exactly what is due the contractor. This amount nearly always varies from the contract price due to additions and deletions from the original agreement.

MOST IMPORTANT FACTOR IN ESTIMATING

It is of prime concern in estimating to include material and labor for every single item required for the job. If something for every item is included, then there is a chance that the effect of mistakes will be reduced by compensating errors. If an item is completely forgotten this will result in a sure loss to the contractor. There have been cases where whole floors have been omitted from an estimate. There should be a complete checklist available for the type of project being estimated. The estimate should be checked by an approximate method, preferably by a second party. Also, if possible, the cost of the project should be compared to a like project already completed. Every reasonable means should be employed to avoid a serious loss due to omission, local foundation conditions, possible injury to adjacent construction, and so on.

THE ANATOMY OF AN ESTIMATE

Most estimates are made up of the following five parts:
1. Materials
2. Labor
3. Equipment
4. Overhead
   a. General overhead
   b. Job overhead
5. Profit

In some cases one or more of the five parts may not apply, as in excavating and in hauling where no raw material has to be furnished, and in some cases where the material or other parts are supplied by the owner.

Materials

The estimator makes a takeoff of all of the different materials required on the job from the plans and specifications. He lists these on quantity sheets according to the applicable units, such as: tons of structural steel, cubic yards of excavation or fill, cubic yards of concrete, mfbm (thousand foot board measure) of lumber, squares (100 square feet) of roofing, linear feet of curb and gutter, thousands of block or brick, etc.

Labor

Considerable judgment on the part of the estimator is necessary in order to correctly forecast the time that will be required to perform each item of the project. Work timetables are a help but, as stated before, the experience of the contractor along this line is the best guide of all; even then the estimator must take into account the influence on work efficiency of local conditions such as: labor supply, wages, union rules, adverse job conditions, etc. There are several methods generally employed to arrive at labor cost; perhaps the best method is to estimate the hours needed to do the required work and then multiply by the appropriate wage.

Equipment

The equipment for a job consists of: all temporary buildings necessary for the job such as toolsheds, storage buildings, and woodworking machinery (including housing for the machinery); material handling machines such as cranes, lifts, earthmovers, and trucks; and small tools. The cost of equipment includes: ownership or rental fees, moving to the job site, erecting, dismantling, operating, etc.

Overhead

In contracting companies large enough to have more than one job going at the same time there are two types of overhead: general overhead and job overhead.

General Overhead. Included in general overhead are all items of expense that cannot be directly charged to any particular job, such as: the cost of maintaining the home office (including office supplies, services, and rent) and the salaries of the office (including office personnel and the principal officers and their travel expenses), promotional costs seeking new business; insurance and taxes; legal and accounting retainers and fees. These costs are prorated to each job.

General overhead will not be significant for the small contractor who maintains no office. For the larger contractor with a number of different jobs the general overhead in most cases varies from about 2 to 8 percent or more of direct costs (materials, labor, and equipment).

Job Overhead. Job overhead costs take care of those items which apply directly to the job and cannot be charged to materials, labor, or equipment and cover such items as: job office expense (including office supplies, rent, services, and personnel); surveys; superintendents, foremen, timekeepers, watchmen, and waterboys; temporary toilets, temporary water, temporary power; barricades and other safety measures; bonds, permits, and inspection costs; payroll taxes and insurance (including Social Security Tax, Federal and State Unemployment Tax, Workmen's Compensation Insurance, Public Liability & Property Damage, etc. Job overhead will vary considerably depending upon the size and type of the job. This figure can vary from 4 to 10% or more of the direct costs.

Total overhead costs may vary from 5 to 25% of the materials, labor, and equipment, or from 10 to 45% of the labor cost depending upon the kind and size of the job and local conditions.

Profit

Most estimators show the profit expected from a job as a percentage of the total estimated cost of the job. This profit customarily varies from 6 to 15% or more depending upon his need for work on the one hand and what he thinks the traffic will bear on the other. Approximate profits usually expected are:

Small jobs	–	15%
Medium jobs	–	12
Large jobs	–	10
Very large jobs	–	6-8

There are times when a contractor will do work at his cost figures in order to keep his key personnel intact and there are times when he will bid high on a job that he does not need, but most of the time his profit figure will reflect the percent which is necessary for a reasonably profitable operation in which he can give the service which will assure him continued work.

SOURCES OF ERROR

There are many avenues through which errors may creep into an estimate:
1. Mistakes in material takeoff
2. Omissions in material takeoff ranging from a few footings or columns to a whole floor
3. Errors in carrying forward material from quantity sheets to summary sheets and from there to the direct cost sheet or sheets
4. Mistakes in estimating the labor time required for certain items of work
5. Errors in estimating hourly wages
6. Failure to allow for rising costs of materials
7. Failure to allow for delays due to breakdowns of machines and acts of God

8. Making no provision to have estimate checked including all extensions and additions
9. Insufficient allowance for overhead
10. Omission of profit

CHECKLISTS

There should be a reminder or checklist for the estimator to use in reviewing his work. This list should be based on similar jobs and cover materials, equipment, and overhead. These lists are helpful but will not supplant a careful study of the plans and specifications.

Sample Checklist

1. Overhead
   a. Main office overhead: part chargeable to job
   b. Job overhead: temporary office and personnel, foremen, watchmen, water boys, sanitary and safety services
   c. Permits: building, street, sidewalk, water, sewer, etc.
   d. Bonds: bid, performance
   e. Insurance: workmen's compensation, public liability, property damage
   f. Taxes: unemployment compensation, social security, sales tax, etc.
   g. Protecting adjacent property
   h. Contingencies
2. Equipment
   a. Wordworking machinery for general carpentry and for the construction of concrete forms
   b. Bending devices for shaping bars for reinforced concrete
   c. Material handling equipment such as: belt conveyors, concrete chutes, elevators, wheelbarrows (both manual and motorized), cranes, trucks
   d. Excavation equipment: bulldozers; power shovels, backhoes, etc.
   e. Pile driving leads and hammers
   f. Compacting devices for fills
   g. Compressors
   h. Small tools (both manual and powered) such as: impact wrenches, saws, tampers, jackhammers, hammers, drills

3. Clearing and grubbing, including wrecking old buildings
4. Excavation

a.	General excavation	e.	Shoring and underpinning
b.	Footing excavation	f.	Sheet piling
c.	Fill and backfill	g.	Well pointing
d.	Blasting	h.	Fine grading

5. Piling

a.	Wood piles	c.	Composite piles
b.	Steel piles	d.	Concrete piles
		e.	Caissons

6. Concrete: footings, columns, beams, walls, floor slabs, roofs, platforms, stairs, steps, etc.
   a. Formwork
   b. Reinforcing steel
   c. Finishes for floors and walls
   d. Special finishes such as terrazzo
   e. Coring
   f. Cold weather concreting
   g. Fireproofing for columns and beams
7. Gravel fill under floor slabs on grade
8. Masonry
   a. Brick: common, face, fire, special
   b. Stone: marble, limestone, sandstone, rubble, random, orchard stone, etc.
   c. Block
   d. Tile
   e. Terra-cotta
   f. Column and beam fireproof covering
   g. Chimneys (cost rises rapidly with height)
9. Precast concrete products
   a. Exterior wall panels
   b. Beams: rectangular, double and single, "Tee Beams," lintels
   c. Columns
   d. Metal panels
   e. Plastic panels
   f. Miscellaneous
10. Waterproofing
    a. Drains around foundations
    b. Membranes on basement walls

c.   Parapet and wall opening flashing
d.   Roof drains and gutters
11. Fascias and soffits
12. Wood construction (light as in houses and offices)
a.   Rough work:

1.	Framing	8.	Rafters	15.	Porches
2.	Studs	9.	Roofing	16.	Stairs
3.	Caps	10.	Subflooring	17.	Rough hardware
4.	Plates	11.	Sheathing	18.	Rough plumbing
5.	Joists	12.	Insulation	19.	Heating
6.	Beams	13.	Furring	20.	Air conditioning
7.	Girders	14.	Bridging	21.	Electrical

b.   Finish work

1.	Siding	18.	Baseboards
2.	Wall shingles	19.	Shoe molds
3.	Roof covering	20.	Stairs
4.	Molding	21.	Railings
5.	Exterior trim	22.	Doors and frames
6.	Interior trim	23.	Windows and frames
7.	Porches	24.	Cabinets
8.	Blinds	25.	Special mill work
9.	Shutters	26.	Mantels
10.	Dormers	27.	Bookshelves
11.	Finish hardware	28.	Plumbing
12.	Flooring	29.	Heating
13.	Floor finishes	30.	Air conditioning
14.	Mosaic tile	31.	Electrical
15.	Vinyl tile	32.	Inside painting
16.	Asbestos	33.	Outside painting
17.	Chair rails	34.	Plastering

13. Prefabricated wood members
a.   Roof decking   c.   Columns
b.   Deep beams   d.   Arches
14. Insulation, walls and ceiling
15. Structural steel
a.   Cost delivered to job:
1.   Base cost of steel plus extras plus freight
2.   Detailing
3.   Fabrication
4.   Transportation to job

b. Erection:
1. Storage at job site
2. Shaking-out
3. Lifting
4. Bolting, welding, or riveting
5. Plumbing
6. Painting
16. Ornamental metals
a. Iron   d. Brass
b. Aluminum   e. Bronze
c. Stainless steel
17. Sheet metal
a. Flashing   d. Roofs
b. Valleys and ridges   e. Downspouts
c. Gutters   f. Prefabricated building partitions
18. Bar joists
19. Wall finishes
a. Lathing: wood, plasterboard, metal
b. Plastering: lime, gypsum, cement stucco
c. Drywall: gypsum wallboard, wood veneer panels
d. Solid paneling: knotty pine, etc.
e. Painted block
f. Brick
g. Blackboards
h. Ceramic tile
i. Wall tile
j. Papering
20. Floor finishes
a. Wood   d. Quarry tile
b. Asphalt tile   e. Ceramic tile
c. Vinyl tile   f. Terrazzo
21. Ceiling finishes
a. Plaster   d. Insulation on slab
b. Hung ceiling   e. Insulation on "Tee" beams
c. Acoustical tile
22. Doors
a. Plate glass entrance doors c. Interior doors
b. Exterior doors   d. Fire doors

23. Windows
    a. Plate glass windows   d. Casement out
    b. Double hung   e. Casement in
    c. Canopy   f. Jalousie
24. Hardware—rough and finish
25. Painting—inside and outside
26. Roof decking
    a. Sonatube   e. Wood deck
    b. Precast concrete slabs f. Concrete
    c. Poured gypsum   g. Miscellaneous
    d. Concrete on metal deck
27. Roofing and flashing
    a. Built-up   d. Copper
    b. Shingle   e. Tin
    c. Roll   f. Tile
28. Plumbing
    a. Rough plumbing: water, gas, sewer lines, trenches, pipe, fittings, valves, etc.
    b. Finish plumbing: toilets, urinals, showers, tubs, lavatories, sinks, disposals, faucets, traps, medicine cabinets, shower curtains, etc.
29. Electrical
    a. Rough: lead-in and all wiring, main fuse box, small fuse boxes, small circuit breakers, transformers, cutout boxes, armored and flexible cable, light and heavy conduit, switches, fixture outlets, both indoor and outdoor receptacles
    b. Finish: fixtures, plates, sockets, outdoor fixtures, time clocks, etc.
30. Heating and air conditioning
    a. Steam and hot water: boiler, stoker, oil tanks, chimney, valves, pumps, piping, fittings, insulation, radiators, heat controls, etc.
    b. Warm air: heating unit, stoker, oil tanks, risers, ducts, insulation, registers, thermostats, etc.
    c. Air conditioning: central or unit system, condenser, mechanical filters, ducts, registers, thermostats, wiring, etc.
31. Fire protection sprinklers
    a. Pumps   d. Valves   g. Fire-hose cabinets
    b. Tanks   e. Sprinkler heads
    c. Pipes and connections f. Standpipes

The Mechanics of Estimating 11

32. Paving
    a. Sidewalks   c. Curb and gutters
    b. Driveways   d. Parking areas
33. Elevators and framing
34. Miscellaneous
    a. Flagpoles
    b. Signs
    c. Vaults

PREPARATION

Before preparing an estimate the estimator should visit the site or have someone else do it for him in order to ascertain any local conditions that would affect the cost of construction, such as:
1. The availability of storage space on the site
2. The cost of off-the-site storage if required
3. Site conditions
4. Local labor conditions
    a. Wage rates
    b. Labor supply
    c. Availability of skilled and common labor
5. Local material supply and delivery costs
.6. Local cost of drayage
7. Availability of electric power, water, etc.
8. Local ordinances and regulations

The estimator should have the cost records of his company at his disposal so that he can know what labor performances he can depend upon. He should also have the delivered prices of the materials which will be included in the project.

He should attempt to familarize himself with the way weather conditions, work delays, and local conditions have affected the overhead costs of his company in the past so that he may have this information as a guide to help him to forecast the probability of the occurrence of such overhead costs in the job he is currently pricing.

It should always be borne in mind that local wage rates and material delivered costs are very important because such costs not only vary from year to year (or in some cases from month to month), but also vary considerably in different sections of the country.

CHAPTER TWO

Procedure

In the remainder of this book we will demonstrate the detailed method of estimating employing the unit quantity procedure. The unit price for each item will include material, labor, and equipment. The items will be summarized and to this total will be added overhead and profit and the cost of a performance bond in order to obtain the bid price.

STEPS IN RECORDING THE ESTIMATE

Each different material or operation of construction such as excavation, concrete work, structural steel, brick, block, woodwork, etc., will be studied. An example estimate of each of these materials in place in the structure will be given. Each of these examples will be accompanied by a set of sketch plans and the following procedure observed.

Quantity Takeoff Sheets

The materials will be taken off the plans shown and placed on the Quantity Takeoff Sheets with the identifying item numbers.

Summary and Unit Cost Sheets

These quantities will be carried forward to the Summary and Unit Cost Sheets and assembled according to item numbers. On these sheets the unit prices will be computed, immediately following each item, with the unit prices for materials, labor, and equipment being kept separate.

Direct Cost Sheets

The item quantities and unit prices will be carried forward to the Direct Cost Sheets and extended to give direct costs. Materials, labor, and equipment costs will appear in separate columns and be summed up in the last or total column.

Overhead and Profit Sheets

Next the general and job overheads will be added on the Overhead and Profit Sheet. General overhead will consist of the job's proportionate part of the main office expense. In job overhead the payroll taxes and insurance will include Social Security, Federal and State Unemployment Compensation, Public Liability and Property Damage, Workmen's Compensation, etc., which will be applied as a percentage of the labor cost. (In the case of subcontracts the subcontractor will absorb these four payroll items and pass them on to the general contractor in the price he bids for his part of the work.)

Then will follow the other job overhead items as discussed in chapter 1. To this will be added an additional amount to allow for unforeseen expenses. This figure will be from 10 to 25% of the computed overhead and will be discussed in more detail later. For instance, if the estimated overhead totaled 8% of the direct costs and an added allowance of 2% was selected, then the total overhead would be 10% of the direct cost.

To the direct cost plus overhead will be added a percentage for profit, which will be determined by factors such as the size of the job, the difficulty of the work, whether the project is hazardous, and whether the contractor has plenty of work or badly needs a job. The percentage can vary from 6 to 15% or more as discussed in the first chapter.

The cost of a performance bond will then be added to obtain the bid price.

ITEM NUMBERS

In recording the information that makes up an estimate it is necessary to use a system of identification to make it easy to tell at a glance to which item an operation or material belongs. This is usually done by means of item numbers (as on Table 2-1).

For instance, we will assign the item numbers from 400 to 499 to concrete. We can start with footings and give the first footing we take off the number 400, under which we will list all the materials, labor, and equipment that are required to complete this footing. All footings of the same type and

TABLE 2-1

Item number	Item	Item number	Item
300	EXCAVATION	1200	PREFABRICATED MATERIALS
400	CONCRETE		1. Precast concrete construction
500	METALS		2. Precast framing
	1. Structural steel		3. Lift slab construction
	2. Bar joists		4. Tilt up concrete panels
	3. Metal decks	1300	SPECIAL EQUIPMENT
	4. Ornamental		1. Blackboards
	5. Miscellaneous		2. Chutes
600	MASONRY		3. Vaults
700	WOOD		4. Revolving doors
	1. Carpentry		5. Kitchen equipment
800	LAMINATED CONSTRUCTION	1400	WALKS AND DRIVES
900	MOISTURE PROTECTION		1. Walks
	1. Waterproofing		2. Drives and parking lots
	2. Roofing		3. Curbs
	3. Flashing	1500	MECHANICAL
	4. Drain tile		1. Heating
1000	GLAZING		2. Cooling
	1. Entrances		3. Plumbing
	2. Doors		4. Electrical
	3. Windows		5. Elevators
	4. Skylights		6. Sprinkler systems
1100	FINISHES		7. Distribution of building costs
	1. Lathing and plastering		8. Average square foot costs
	2. Dry wall construction	1600	GENERAL OVERHEAD
	3. Resilient floor tile	1700	JOB OVERHEAD
	4. Ceramic floor and wall tile	1800	PROFIT
	5. Terrazzo		
	6. Painting		

Note: It should be noted that the item numbers through 1500 are keyed to the chapters in which each item is discussed. For instance, concrete is considered in chapter 4 and therefore the item numbers for concrete construction range from 400 to 499. The items of General Overhead, Job Overhead, and Profit are an exception to this rule for two reasons: (1) since they are applied at the end of the esti-mate they are given the numbers following the last (or 15th) chapter—these numbers being 1600, 1700, and 1800; and (2) since these items must be used in making example estimates they must be discussed before these estimates are presented. Therefore, they are discussed in chapter 2 where the procedure for preparing an estimate is developed.

same general size will be given the same number. Each different type of footing will get another number such as 401 or 402. When we think that we have covered all of the footings and end up with, say, item 407, we will leave several numbers vacant for footings that we may discover later. Then the next item, which will be slabs, will start with number 410 or 412 and so on.

Table 2-1 gives the item numbers which we will use in the remainder of the text and which should be observed in the homework problems.

EXAMPLES

After each example given in the text there will be a practice example with sketch plans which the student should complete using Quantity Takeoff Sheets, Summary and Unit Cost Sheets, Direct Cost Sheets, and Overhead and Profit Sheets. He should use the illustrated example as a guide. A blank copy of each of these forms is given in Appendix F (pp. 286-289).

After the student has completed all of the examples covered in the text, he should be able to make an intelligent estimate from a set of building plans. If time permits it is recommended that the student be required to complete an estimate of a building in the $500,000 to $1,000,000 class designed for the area of the country in which he lives.

PRECISION OF EXTENSIONS

In computing the unit prices for materials, labor, and equipment the amounts should be carried to the nearest cent; but when extending the prices on the Direct Cost Sheets to get the total cost of an item the amount should be rounded off to the nearest dollar. Some of the prices will go up to the higher dollar and some will drop to the next lower dollar and will thus very nearly balance out to the same total as if the amount for each item were carried to the nearest penny.

WAGES

The wages that will be used in all examples are the average wages for the 30 largest U.S. cities for 1971 and include fringe benefits but no insurance or taxes. A table of these wages is shown in Appendix B.

PAYROLL TAXES

As stated before there are two payroll taxes whose cost must be added to every job: the Social Security Tax and the Federal and State Unemployment Compensation taxes. In each case the amount of the tax is computed as a percentage of the direct labor wages.

Social Security Tax

The Social Security Tax, as of 1973, will cost the contractor 5.85% of the first $10,800 of wages paid every year.

Federal and State Unemployment Tax

The Federal Unemployment Tax rate, as of 1972, is 3.2% of the first $4,200 of wages paid during each year. Every state has an unemployment tax and no two of them are exactly alike. The federal government collects this tax for the states and credits the federal tax due with the amount of the state tax up to 2.7%.

The State Unemployment Tax rate is 2.7% except for the following 8 states:

Alaska	—	2.9%	New Jersey	—	2.8%
Hawaii	—	3.0	North Dakota	—	4.2
Idaho	—	2.7 to 3.3	Ohio	—	4.0
Nevada	—	3.0	South Dakota	—	3.6

In all but 18 of the states the tax is figured on the first $4,200 of wages just as is the Federal Tax. The tax base for these 18 states is as follows:

Alaska	— $7,200	Massachusetts	— $3,600	Rhode Island	— $3,600
Arizona	— 3,600	Michigan	— 3,600	Tennessee	— 3,300
California	— 4,100	Minnesota	— 4,800	Utah	— 4,300
Delaware	— 3,800	Nevada	— 3,800	Vermont	— 3,600
Hawaii	— 4,300	Oregon	— 3,600	West Virginia	— 3,600
Idaho	— 3,600	Pennsylvania	— 3,600	Wisconsin	— 3,600

These state tax rates are in most states the standard rate—that is, the rate required of employers until they are qualified for a rate based on their experience record for unemployment.

Since all of the states have different rules and restrictions and since every contractor has a different experience record the exact tax rate for each particular contractor in each state in which he works must be determined separately.

In the examples in this text we will use the state rate of 2.7% for the first $3,000 of wages paid. As already noted the federal rate is set by statute at 3.2% of the first $4,200 of wages paid.

INSURANCE

As mentioned before the contractor must carry insurance to cover the cost of Workmen's Compensation and also to provide for Public Liability and Property Damage. The cost of these two insurance coverages is in each case based on a percentage of the direct labor wages.

Workmen's Compensation

Each state has its own table of "Manual" rates covering all of the different types of work subject to the provisions of the Workmen's Compensation Act for that state. These are the basic rates per $100 of payroll applicable to each classification.

For the contractor whose premium payment per year is large enough to qualify for an experience rating (usually over $750) these "Manual" rates will be adjusted according to the contractor's own safety experience record. In Nevada, North Dakota, Ohio, Oregon, Washington, West Virginia, and Wyoming, the Workmen's Compensation coverage is handled through state financed organizations in place of private insurance carriers.

Public Liability and Property Damage

As in the case of Workmen's Compensation, there are also basic "Manual" rates for Public Liability and Property Damage insurance that differ in each state, and there are also similar limitations in regard to the application of payroll between different insurance classifications. The experience of the contractor also has a bearing on the adjusted rates.

Tables for the manual rates that apply to building contracting for Workmen's Compensation and Public Liability will be shown in Appendixes C and D. (Some states have insurance coverage requirements in addition to those listed here and the estimator should familiarize himself with any such exceptions.)

In the examples in this text we will use an average rate for the cost of Workmen's Compensation and Public Liability and Property Damage. In actual practice these rates must be determined for each contractor in each state in which he works.

PERFORMANCE BONDS

The approximate cost of performance bonds for building work, based on the amount of the contract, is:

For the first	$ 100,000	—	$10.00 per $1,000
For the next	2,400,000	—	6.50 per 1,000
For the next	2,500,000	—	5.25 per 1,000
For the next	2,500,000	—	5.00 per 1,000
For all over	7,500,000	—	4.70 per 1,000

GENERAL OVERHEAD ITEMS

For the purpose of developing the overhead and profit philosophy let us examine a contracting firm which is a two-man partnership doing an average business of $3,000,000 per year with both partners being 40 years of age.

First consider the general overhead for a firm this size. There will probably be a small office with an adjoining equipment yard and storage sheds on the outskirts of the city which will be valued at, say, $75,000. Then there will be the cost per year of utilities, taxes, insurance, upkeep, salaries, etc. To arrive at the total cost of general overhead per year assume:

Interest cost for office and equipment yard $75,000 @ 9%	$6,750
Telephone, water, lights, etc.	1,500
Office supplies	300
Upkeep	450
Taxes and Insurance	2,000
Secretary—Salary	6,000
General supt.—Salary	12,000
2 Partners—Salaries	36,000
Legal and accounting fees	1,000
Travel	5,000
Total General Overhead for Year	$71,000

Assuming 10% for general plus job overhead and 10% for profit, the direct cost to produce $3,000,000 worth of contracts would be $3,000,000/(1.1 × 1.1) = $2,479.340. Then the percentage of direct cost to allow for general overhead would be $71,000/$2,479.340 = 2.86%. This cost will usually vary from 1½ to 3%.

When all of the expected job overhead costs that are listed in chapter 1 have been established, then something should be added to cover costs which cannot always be predicted. This extra cost will not be large if the overhead has been carefully considered. However, to guard against this eventuality an amount should be added to the computed job overhead figure in the range of 10 to 25% of this computed job overhead.

As an example, suppose that for a particular project the direct cost (labor, materials, and equipment) was estimated to be $500,000, with general overhead entered as 2% and job overhead, for the usual expected costs, as 6%. The cost of the project would be:

Direct cost	$500,000
Overhead — 2% + 6%	40,000
	$540,000
Profit — 10%	54,000
	$594,000

The job was completed and the books showed a profit of 7.4% instead of the 10% sought. In reviewing the construction it was noted that there were two items of expense for which nothing had been provided in the overhead forecast: some concrete beams and part of a floor slab had to be broken out and replaced at a cost of $5,000; and a supply item had advanced $5,000 in cost. When this extra cost of $10,000 was subtracted from the $54,000 profit the percentage of profit became ($54,000 - $10,000)/594,00 = 7.4%.

These were unusual happenings and could not be foreseen. Now if 25% had been added to the computed overhead of 8% (made up of 2% general plus 6% job), then the total overhead in the bid would have been 10% and the bid price necessary to give a 10% profit would have been:

Direct cost	$500,000
Overhead — 10%	50,000
	$550,000
Profit — 10%	55,000
	$605,000

In addition to these costs there is another item due to the fact that 10% of the cost of the job is withheld each month by the owner to be paid in one lump sum at the completion of the project and its acceptance. Because of this the contractor must finance 10% of the work himself for the duration of the contract. Assuming an interest charge of 9% this would mean $0.09 \times 0.10 = 0.009 = 0.9\%$ of direct costs for this item ($0.9\% \times$ direct cost $\times$ years in job $\times 0.5$).

There will be times when collections will not be sufficient to meet all of the payrolls and to take advantage of the discounts given for the payment of bills before the tenth of the month. Money will have to be borrowed for this purpose. The amount of this borrowing is difficult to estimate beforehand and so this item falls into the category of unexpected overhead. It is not usually of significant size and is pretty well offset by the amount of earned discounts.

There is another item of job overhead that stems from equipment. It often happens that some pieces of equipment are kept on the job and used to handle several different materials. The cost of this type of equipment can be more easily estimated as job overhead rather than trying to divide the cost between the various items affected.

The equipment could include:

1. A pickup truck which is used by the job superintendent and will carry no labor cost since he will use it as his private transportation in connection with the work and will do the driving himself.
2. One or more trucks to handle miscellaneous hauling during the job. The cost will include rentals, operating costs, maintenance, and labor.
3. Lifting equipment which is used during the different phases of the work and cannot be charged to any one item exclusively and which will carry rentals, operating costs, maintenance, and labor. This equipment will include such things as:
    a. Stationary hoists to lift materials to the upper floors.
    b. Hammerhead cranes that are left in place until the job is completed.
    c. One or more movable cranes, caterpillar or truck.

There is one more item of job overhead that needs careful scrutiny. It should be kept in mind while computing the Social Security Tax that this levy applies only to the first $10,800 of the annual earnings of the individual. To allow for this fact the 5.85% rate should be applied to only a part of the total wages paid. In the case of Federal and State Unemployment taxes the rate applies to the first $4,200 of annual earnings of the individual. To allow for this fact the rate of 3.2% (which we will use in our examples) should also be applied to only a part of the total wages paid.

Unlike office workers who will be on the job most of the available working days the construction worker will miss many days in the year due to rain, snow, cold weather, and wind storms. In many cases this will bring his yearly wage below the maximum amount to which the tax applies so that all of his wages will be subject to the tax. Then as his yearly income increases, above the maximum figure, less than his total income will be subject to the tax.

In construction work certain trades will lose more time due to weather conditions than others. For instance, steel workers will not put up steel if even a few drops of rain or a little snow falls because this will make the steel slippery and dangerous to handle and walk upon. Concrete cannot usually be poured when the temperature is below 30 degrees or in the rain and neither can brick be laid in like conditions. Of course, when the building is framed and roofed then the inside finish can be accomplished with only a few days lost due to extreme cold or other very inclement weather.

As an average figure for the whole job about 20 days per year may be lost in the South, perhaps 60 days or more in the far North, and around 40 days in those areas between the South and North.

For workers without paid vacations, which will include almost all construction workers, the available work days per year will be approximately: 365 minus 52 Saturdays, 52 Sundays, and 6 holidays, making 255 working days.

Labor turnover will also affect the amount of wages subject to taxes because wages paid to a worker who works for the contractor less than a year will hardly ever be equal to the maximum taxable amount and therefore all of his wages would be subject to the tax.

An average percentage of wages subject to Social Security and Federal and State Unemployment taxes may be determined as follows:

Assume 40 days lost from the available work days in a year. The remaining time will be 255 minus 40 equals 215 days or 1,720 hours.

For an average of $7.00 per hour the yearly wage will be $7.00 x 1,720, which equals $12,040. With a 15% labor turnover, then three out of every twenty work positions will be temporary. The amount subject to Social Security would be (using the $10,800 limit);

17 workers @ $10,800	$183,600
3 temporary positions @ $12,040	$36,120
Total	$214,500

The percentage of wages paid which would be subject to the tax would then be $214,500 divided by (20 x $12,040), equaling 89%. Use 90% since the figure is approximate anyway.

For Social Security Tax use the following percentages of wages paid:

For a $6.00 average hourly wage—100%
For a $7.00 average hourly wage— 90%
For a $8.00 average hourly wage— 80%
For an $9.00 average hourly wage—75%

For Federal and State Unemployment Tax, when the amount subject to the tax is $4,200, then the percentage of wages which would be subject to the tax would be:

For a $4.00 average hourly wage—65%
For a $5.00 average hourly wage—57%
For a $6.00 average hourly wage—50%
For a $7.00 average hourly wage—45%
For an $8.00 average hourly wage—40%

PROFIT PHILOSOPHY

As discussed earlier there are many things that will affect the profit which a contractor will add to his estimate. This amount will vary with the size of the contract, the type of work, the time required for completion, whether or not the work is hazardous, whether he has plenty of work, or whether he needs work to keep his key personnel busy. Nevertheless, there is an average overall profit that a firm needs to earn in order to realize its financial goals.

In addition to their salaries, which are a part of the general overhead, the partners hope to clear enough profit to:

Pay themselves bonuses—say 2 @ $38,000	$76,000
Pay bonuses to key employees	13,197
Arrange retirement pay for 4 key personnel whose average age is 35 years—($500 per month each)	9,100
Provide life insurance policies on each partner payable to the other so that he will have the cash to buy out his partner's interest at his death for $150,000 (policy also to cover $1,500 per month retirement for each partner, both partners being 40 years of age at time of writing of policy)	17,430
Set aside for working capital and to purchase securities to serve as collateral	30,000
North Carolina state income tax on profit* of $272,727 would be approximately 2 × $6,000	12,000
Federal income tax would be approximately 2 × $50,000	100,000
Promotional activities, etc.	15,000
Total	$272,727

*10% profit on work bid in at $3,000,000 would be

$3,000,000 ÷ $\dfrac{.1}{1.1}$ = $272,727.

So we see that in the case of this company an average profit of 10% would be needed to accomplish the stated objectives.

The preceding discussion was based on a partnership relationship. In actual practice when any business reaches this size it is organized as a corporation for several reasons:

1. Personal liability to outsiders and creditors is limited to the extent of the personal investment in the corporation.

2. Valuation of personal interest in business at time of death of one or both partners is greatly simplified and facilitated by "pegged" value of shares of stock. This saves the type of dispute that arises in a partnership where an amount has to be agreed upon for the "good will" value of the business.

3. There are some tax advantages also.

However, the method of estimating the amount of profit a firm will have to make to realize its financial goals will be the same and the percentage of profit required will be approximately the same.

CHAPTER THREE

Excavation

Before excavation can begin it is necessary to remove trees, stumps, brush, etc. The cost of clearing and grubbing by bulldozer or tractor varies from about $250 to $750 per acre depending upon the size and type of trees to be removed. Removing trees by hand is very expensive; for instance, the cost of removing trees and stumps by hand will vary from $25 to $200 each depending upon the difficulty of the work.

Before excavation and grading are done the topsoil is usually removed and stored to be used later for landscaping. This excavation is classed as general and can be done by equipment employing a blade. Spreading the topsoil back in place after the building is completed is usually done under a subcontract for landscaping.

TYPES OF EXCAVATION

There are two types of excavation:

1. *General excavation* refers to the removal of large quantities of material, such as railway or highway fills and large basements. This type of excavation is usually done by power shovels, bulldozers, scrapers, etc.

2. *Structure excavation* refers to the digging of holes for footings, trenches for pipe, etc. At a building site the excavation needed to bring area to grade outside of the building limits would be general. Also, if the first floor were on grade, the excavation to the bottom of the gravel under the slab would be classed as general since it too could be done with a blade. The remainder of excavation, below this elevation, for footings and so on, would be classed as structural and done with a clamshell or backhoe and in some cases by hand where it cannot be reached by machine.

CLASSES OF SOIL

There are four principal classes of soils:

1. *Light and Medium Soils,* which can be moved by blade, clamshells, shovels, backhoes, etc.
2. *Very Hard Soils* such as stiff clay and compacted loams mixed with gravel. This type of material requires heavy-duty shovels for digging or, if lighter equipment is used, then the material must be loosened in some manner.
3. *Shales or Hard Pan,* which require at least partial blasting before they can be moved by power equipment.
4. *Rock,* which requires considerable blasting before it can be removed.

SHRINKAGE AND SWELL

Almost all types of soils and also rock will increase in volume when excavated. This is called "swell" and can amount to from 10 to 40% and sometimes a little more. Thus, a truck of 4-cu-yd capacity will haul only 4/1.20 or 3.33 cu yd of bank-measured material having a swell of 20%. On the other hand, soil placed in a fill or trench can be compacted to less than its bank measure and shrink by as much as 10 to 15%.

SHORING

Some types of soils will stand with a vertical face, while others will require shoring to hold back the unexcavated portions or the bank will have to be sloped to the angle of repose of the soil.

EXCAVATIONS NEXT TO AN ADJOINING BUILDING

Excavations next to an adjoining building have to be carried on with care not to disturb the foundation of the existing building. This can be done by shoring up the bank as the excavation advances or by sheeting driven next to the existing building and braced as the excavation proceeds.

EXCAVATIONS ON WET SITES

When a bridge pier is to be constructed in the bed of a stream two coffer-dams are built around the pier site with a space between them which is filled with earth. The water is pumped out of the inner enclosed space, thus allowing the work to proceed in the "dry."

When a building foundation has to be placed below the water table, well points are sometimes used to draw down the water elevation over the entire footing area. A well point is a perforated pipe with the openings protected by a screen. It is jetted into the ground so that the lower end is below the elevation to which the water table must be drawn. The well points are usually spaced from 2 to 5 feet around the perimeter of the area involved and connected at the top to a header pipe which is usually 6 to 10 inches in diameter. The water seeps into the well points and is pumped out through the header usually by a self-priming centrifugal pump.

In soil that is not very permeable a column of coarse sand or fine gravel is placed around the pipe to collect the water. This is done by first driving a larger pipe into the ground, removing the soil from the inside of the pipe, installing the well point, placing the sand or gravel in the space between the two pipes, and withdrawing the pipe sleeve. (An expert in this line should be consulted in making an estimate for dewatering any site.)

COMPUTING EXCAVATION QUANTITIES

When an excavation is to be made elevations are usually taken at the break points on the existing ground and contour lines plotted at one- or two-foot intervals. Then, after the proposed shape of the ground surface is determined, a new set of contour lines is plotted to suit this new condition using the same interval as above. Both sets of contour lines are shown in the same sketch—one solid and the other dotted or dashed. The simplest way to determine the amount of excavation is to cut sections at representative points and plot both the existing and proposed contours in each section joining each set of coordinates with a straight line. The area of each section is determined and the volume computed by averaging each two adjacent areas and multiplying the result by the distance between the two sections. The result should be in cubic yard units. If the sections are plotted on graph paper, the areas can be determined by counting the squares between the curves. It is relatively simple to compute the areas using the plotting coordinates of each point as shown in Fig. 3-1.

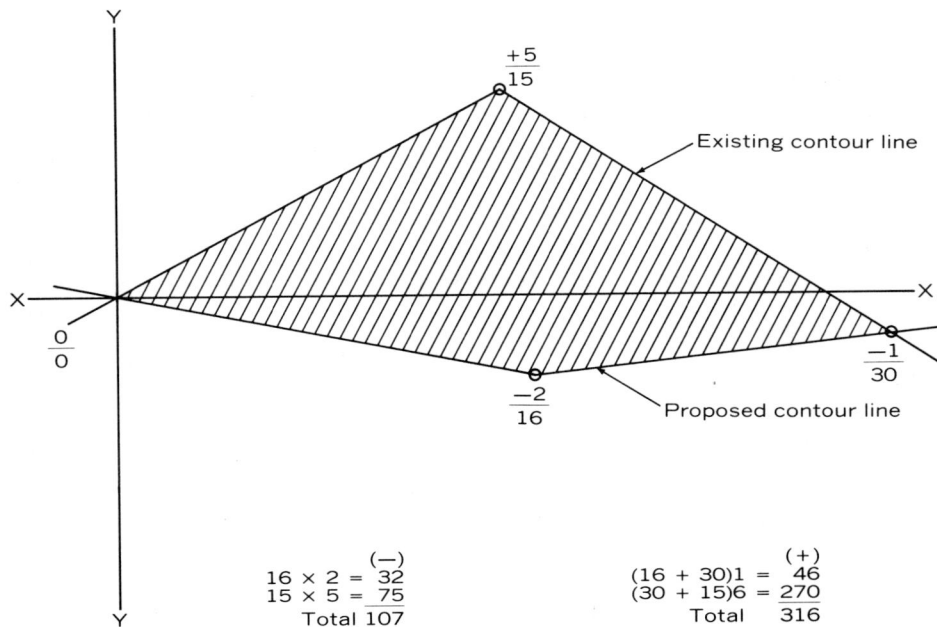

Existing contour line

Proposed contour line

	(−)		(+)
16 × 2 =	32	(16 + 30)1 =	46
15 × 5 =	75	(30 + 15)6 =	270
Total	107	Total	316

Area = (316 − 107) ÷ 2 = 104.5 sq. ft.

FIGURE 3-1
Method of Computing Areas Between Contour Lines

Procedure for computing volumes of cut and fill by taking sections through contour lines:

In Fig. 3-1 the top figure of the coordinate of a point shows the number of feet above or below the *X-X* axis; the bottom figure shows the distance out from the *Y-Y* axis as +5/15, which means that the point is 5 ft above the *X-X* axis and 15 ft to right of the *Y-Y* axis.

In this solution we will progress in a counterclockwise direction around the figure in this manner: using the first two sets of coordinates, 0/0 and -2/16, we obtain the algebraic difference between the top figures, 0 - (-2) = 2, which is the difference in elevation between the two points, and multiply by the sum of the bottom figures, 0 + 16 = 16. This will be 16 x 2 = 32, which will be in the minus column because the line between the two points is descending and therefore the area must be subtracted.

Take coordinates, − 2/16 and − 1/30 = [− 2 − (−1)] (16 + 30) = 1 × 46 in plus* col.

Take coordinates, −1/30 and +5/15 = [−1 − (+5)] (30 + 15) = 6 × 45 in plus* col.

Take coordinates, +5/15 and 0/0 = [+5 − (0)] (15+0) = 5 × 15 in minus col.

The total is (+316 − 107)/2 (because we have doubled the area) = 104.5 sq ft.

Note: In computing volumes when the interval between the chosen sections is constant add each two adjacent areas and multiply by the interval ÷ (2 × 27) in order to get cubic yards. For example: for 50′ intervals, the multiplier is 50 ÷ (2 × 27) = 0.926; for 25′ intervals, the multiplier is 25 ÷ (2 × 27) = 0.463.

As an example, for six areas, A1, A2, A3, A4, A5, and A6, @ 50′ intervals: Volume = [A1 + 2(A2 + A3 + A4 + A5) + A6] × 0.926 = (Answer in cu yd)

EXCAVATING AND BACKFILLING BY HAND LABOR

Excavating and backfilling by hand labor is very seldom, if ever, done except in the rare cases where excavating equipment cannot be used for one reason or another and in the trimming out process where equipment cannot be used, such as in inaccessible corners.

The cost of doing this work by hand varies with the hardness of the soil and the depth of the excavation. With hard soil a lot of work with a pick may be needed to loosen the material before the actual moving of the soil can begin. Also, if the depth of the excavation is over 5 feet then the material will have to be handled more than once, which will raise the cost. Hand excavation and backfilling can rarely be done for less than $6.00 per cubic yard and can go much higher, for instance, when the excavation is over 5 feet deep, and loosening by pick is required or bracing is needed.

---

*This sign is plus because the line is ascending in a counterclockwise manner.

The cost of excavation will vary depending on: whether the material is loose soil which can be easily dug; whether it is hard soil which must be broken up before it can be handled; whether it is very hard soil or rock which requires blasting; whether the sides will remain vertical or will have to be braced; whether the site is wet or dry; what will have to be done with the excavated material; how much backfilling and compaction will be required; the type of equipment that will be employed; working conditions and other miscellaneous factors.

All heavy equipment such as trucks, cranes, shovels, scrapers, backhoes, compressors, etc., will be estimated in this book as rented equipment since for most small contractors it is cheaper to rent a piece of equipment while needed rather than to have it sit around idle while not being used. The rental figures quoted in this text are the national average as compiled during the latter part of 1970. The age and condition of the equipment are taken into account as well as the variations due to the different regions where lengths of working seasons, climatic factors, local practices, and soil conditions assert their influence. Therefore these figures can only serve as guides; and even these average figures would have to be substantially increased should the equipment be perfectly new. A large general contractor will maintain an equipment pool and keep all of the equipment in use most of the time by moving it from one job to another and only bringing it to the equipment yard for repairs or for servicing. Each such concern has its own special cost experience for its machines, which will of course be available to the estimator. Large contractors often find it cheaper to own their equipment, yet will, in some cases, use rentals. In this text we will deal only with medium-size jobs and small- to medium-size contractors.

However, we will discuss equipment ownership. Appendix E will show excerpts from a set of tables which is a compilation of data on the contractor's average costs of owning, maintaining, and operating construction equipment. This material is being shown by courtesy of The Associated General Contractors of America with this caution, "This book of tables is intended as a guide only and all average percentage rates carried therein are subject to adjustment to fit the experience of the individual equipment owner."

The items of equipment ownership expense are expressed as a percentage of the value of the equipment. This value should be the replacement cost since in the present era prices are fluctuating considerably every few months.

The general procedure for setting up the cost of the excavation necessary for the building project is as follows:

1. Compute the required number of cubic yards of excavation.
2. Select equipment to be used.
3. Determine output per hour of equipment.
4. Find rental cost per hour of selected equipment including cost of operation.
5. Ascertain hourly rate for operating crew.
6. Compute the cost per cubic yard of excavation for: (a) labor and (b) equipment.

EXAMPLE 3-1

Compute the labor and equipment costs for the general excavation for a one-story classroom building 200 ft by 160 ft out to out of walls. It will be slab on grade with a 5-in. concrete slab on a 6-in. gravel fill. The average original contour elevation at building will be 2 ft and the finished grade at bottom of gravel fill will be 0 ft. (See Fig. 3-2.) The column footings will be ignored since they are not general excavation. The entire campus will cover 10 acres and there will be 4,000 cu yd of excavation outside of the building to be made. All excavation will be used to bring the area to finished contour

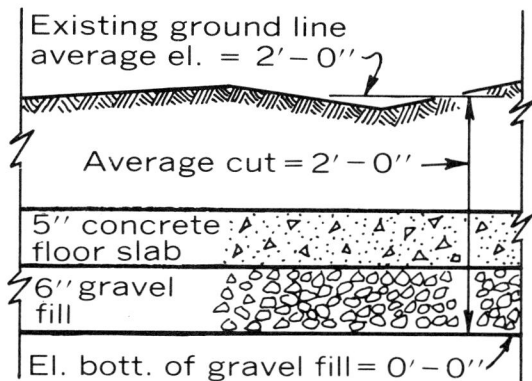

FIGURE 3-2
Typical Section Through Floor

elevations. The top soil will be moved to one side and respread later under the contract for landscaping. The soil will be medium clay. Only a medium stand of timber exists on 3 of the acres: about 20 trees to the acre of 8-in. to 12-in. diameters. An allowance of $300 per acre for the 3 acres will take care of clearing, grubbing, and burning. The six steps required to prepare this estimate are as follows:

1. The required excavation is $(200 \times 160 \times 2)/27 = 2{,}370$ cu yd in building plus 4,000-cu yd area excavation equals 6,370-cu yd total. This will be shown on quantity sheets and the total carried to the summary sheets.

2. Select a unit composed of a diesel-powered wheel-type tractor of 170-brake horsepower and a self-loading wheel-type scraper having 12-cu yd heaped capacity from Table 3-1.

TABLE 3-1   Output per Hour in Cubic Yards

Haul distance in feet each way	400	750	1,000	1,400	2,400	4,000
Wheel-type tractor, 170 hp (and wheel-type scraper) Load—bank measure 12 cu yd heaped = 10 cu yd	125	105	100	80	65	55
Crawler-type tractor, 120 hp (and wheel-type scraper) Load—bank measure 12 cu yd heaped = 10 cu yd	95	70	60	45	30	20

Up to hauls of perhaps 600 feet crawler-tractor-pulled scrapers are used because of high drawbar pull and better traction, but for longer hauls the speed of the wheel-tractor-scraper unit makes it more economical.

Also select a 115-hp diesel engine bulldozer crawler-type tractor to help load the scraper.

3. To compute output per hour assume: haul distance of 600 feet; that material will swell 20%; that equipment will work 45 minutes out of each hour; that due to swell net load will be 12/1.20 = 10 cu yd bank measure; speed during loading will be 1 mile per hour or 88 feet per

minute, and scraper will load in distance of 150 feet; speed moving will be 8 miles per hour or 704 feet per minute.

Time for each trip will be:

Loading 150/88   = 1.71 min
Hauling (600/704)2   = 1.70 (round trip)
Dumping   = 0.75
Maneuvering   = 0.25
Total time per trip   = 4.41 min
Trips per hour   = 45/4.41 = 10.2
Volume per hour = 10.2 x 10 = 102 cu yd

Table 3-1 shows the approximate output for one size of wheel-type and crawler-type tractor scrapers, in cubic yards, per hour, bank measure with 20% swell of soil on fairly level terrain in medium earth.

4. Rental cost of scraper and hauling unit based on a weekly rental rate is 650/40* hours = $16.25 per hour. Operating costs, which include fuel, oil, grease, filters, etc., are (see Note for method used):

Cost of diesel fuel = $0.30 x 2/3 x $0.55   = $0.011
Cost of oil, grease, filters, etc. = 0.55 x $0.011 =   0.006
Cost per rated hp   = $0.017
Operating cost per hour = $0.017 x 170 hp=$2.89
Rental plus operating cost per hour = $16.25 + $2.89 = $19.14

Note: An approximate method of computing the cost of operating equipment powered by diesel engines is as follows:

Assuming cost of diesel fuel at $0.30 per gal
Assuming fuel consumption at 0.055 gal
  per brake hp hour
Then cost of fuel = 2/3 x 0.055 x $0.30   = $0.011
Cost of oil, grease, filters, etc.,
  55% of fuel cost   =   0.006
Total cost per rated hp hour   = $0.017

[*Note is continued on next page.*]

*From Appendix A (where not shown figure weekly rental rate 1/3 of monthly rate).

When the equipment is powered by gasoline engines the cost is as follows:

Assuming cost of gasoline at $0.40 per gal
Assuming fuel consumption at 0.06 gal
  per brake hp hour
Assuming equipment operates at 2/3
  rated hp
Then cost of fuel = $0.40 × 2/3 × 0.06 = $0.016
Cost of oil, grease, filters, etc.
  = 0.4% of fuel cost   =   0.006
Total cost per rated hp hour   = $0.022

A 115-hp diesel engine bulldozer crawler-type tractor will be used to help load the scraper and shape the fill. Rental cost including blade based on a weekly rental basis is $551/40 hours = $13.78 per hour. Operating cost will be $0.017 × 115 hp = $1.96 per hour. Rental plus operating cost per hour = $13.78 + $1.96 = $15.74.

5. Hourly rates for the two operators will be computed at $8.20 per hour each; one foreman will be listed at $8.55 per hour.

6. Cost per cu yd of excavation is found as follows:

a. Labor cost per cu yd:

$$\frac{\text{operators' wages per hour}}{\text{cu yd moved per hour}} = \frac{(2 \times \$8.20) + \$8.55}{102} = \$0.25$$

b. Equipment plus operating cost per cu yd:

$$\frac{\text{total rental plus operating cost per hr}}{\text{cu yd moved per hour}} = \frac{\$19.14 + \$15.74}{102} = \$0.34$$

The quantities and prices can now be carried to direct cost sheets and extended. Then overhead and profit can be added.

EXAMPLE 3-2

Compute the labor and equipment costs for digging a basement 200′ × 160′ × 8′ deep in medium clay soil and hauling 3 miles in dump trucks. After the basement is excavated the following holes must be dug for the column footings: 18 holes 12′ × 12′ × 6′; 21 holes 9′ × 9′ × 5′; and 8 holes 7′ × 7′ × 5′. The dirt will swell 20%. Trucks will travel 25 mph loaded and 30 mph empty.

Using the same procedure as in Example 3-1 consider the basement first.

PART 1

1. The required excavation for basement is (200 × 160 × 8) / 27 = 9,480 cu yd. This computation will be shown on quantity sheets and the total carried to summary sheets.
2. Select a diesel-powered crawler shovel with a 2½-cu-yd-bucket and 12-cu-yd-capacity dump trucks. See Table 3-2.

TABLE 3-2
Approximate Maximum Bucket Capacities per Hour
in Cubic Yards

Bucket size	3/8	1/2	3/4	1	1 1/4	1 1/2	1 3/4	2	2 1/2
Optimum depth	4.4′	5.5′	6.5′	7.5′	8.5′	9.0′	9.5′	10.′	11.′
Common earth	70	90	130	170	200	230	260	300	350
Optimum depth	6.′	7.′	8.′	9.′	10.′	10.8′	11.5′	12.′	13.′
Hard clay	50	70	100	140	170	200	225	260	325
Wet clay	25	35	65	90	100	150	160	180	210

3. To compute output per hour: Truck capacity in bank measure = 12/1.20 = 10 cu yd. From Table 3-2 approximate bucket capacity is 300 cu yd per hour or 300/60 = 5 cu yd per minute. (Bucket is digging at less than optimum depth.)

Truck time: To maneuver and load	=	2.5 min.
To travel loaded		
= (60 min/25 mph) × 3 mi	=	7.2
To return empty		
= (60 min/30 mph) × 3mi	=	6.0
Delays and dump time	=	4.3
Total trip time		20.0 min.

Each truck can then make 60/20 or 3 trips per hour.

Shovel can load each truck in 10/5 = 2 minutes.

Since it takes an extra half minute to get into position for loading, the number of trucks that can be loaded and moved away per hour = 60/ 2.5 = 24. Since each truck can make only 3 trips per hour, then the number of trucks required to keep shovel busy will be 24/3 = 8 trucks. Assume gang of foreman, operator, 8 drivers, 2 laborers.

Table 3-2 shows the approximate maximum bucket capacities in cubic yards, bank measure, for 60 minutes per hour operation with the given optimum depths (optimum depth meaning that depth of cut at which the particular size of bucket operates most efficiently) with angle of swing between 45° and 90°.

Note: All of the tables, wages, and overhead percentage items given in this text are approximate averages and are not to be used in estimating actual work in any given locality. Prices and equipment performance vary and should be determined for the area in which the work is to be accomplished.

4. Rental cost of shovel based on a weekly rental rate is $1640/40 = $41.00 per hour. Operating cost (fuel, oil, grease, etc.) will be $0.017 per brake hp × 175 hp (assumed) = $2.96. Rental plus operating cost per hour = $41.00 + 2.96 = $43.96. (See Appendix A—proportion for 2 1/2-cu yd bucket and assume weekly rental 1/3 of monthly rate.)

   Rental cost of 12-cu-yd-capacity diesel dump truck based on a weekly rental rate* is $490/40 = $12.25 plus allowance for tire wear† of, say, 0.34 = $12.59. Operating cost will be $0.017 per brake hp × 185 hp (assumed) = $3.14. Rental plus operating cost per hour per truck will be $12.59 + $3.14 = $15.73.

5. Hourly rate for shovel operator   = $8.20
   Hourly rate for each truck driver   = 6.20
   Hourly rate each laborer   = 6.05
   Hourly rate for foreman   = 8.55

---

*From Appendix A (where not shown figure weekly rental 1/3 of monthly rate).
†$1,500 (tires) / 6,250 hr (life) = $0.24 + (Repairs 0.10) = $0.34/hr

6. Cost per cu yd of excavation then becomes:

    a.   Labor cost per cu yd

$$\frac{\text{operators wages/hr}}{\text{cu yd moved/hr}} = \frac{\$8.20 + (8 \times \$6.20) + (2 \times \$6.05 + \$8.55}{8 \text{ trucks} \times 3 \text{ trips / hr} \times 10 \text{ cu yd}} = \$0.33$$

    b.   Equipment plus operating cost per cu yd

$$\frac{\text{total rental plus operating cost/hr}}{\text{cu yd moved/hr}} = \frac{\$43.96 + (8 \times \$15.73)}{8 \times 3 \times 10} = \$0.71$$

For a 5-mile haul 12 trucks would be required to keep the shovel busy and move 240 cu yd per hour (the same amount that 8 trucks moved with the shorter haul) and the cost would be $0.43 for labor and $0.97 for equipment.

PART 2

Now the column footing excavation will be priced.

1. The required excavation will be 18 $(12' \times 12' \times 6') + 21 (9' \times 9' \times 5') + 8 (7' \times 7' \times 5')/27 = 963$ cu yd

2. Select 3/4-cu yd diesel-engined backhoe.

The backhoe is excellent for digging trenches, footings, and basements because it can dig vertical walls in soil which will stand vertical; it can cut corners square; and it can trim the floor level. The machine is operated from the bank and is not affected by a soft or wet hole in which it is digging. Very little hand trim is required with a backhoe. Also select two 10-cu-yd-diesel-engined dump trucks to move excavation to storage pile.

3. To compute output per hour:

Truck capacity in bank measure = 10/1.20 = 8.33 cu yd

Backhoes may be considered to be able to dig about 2/3 as much as the same size power shovel. So a 3/4-cu-yd backhoe will be assumed to dig $2/3 \times 75$ per hour or 50/60 = 0.833 cu yd per minute.

(In Table 3-2 for 3/4-cu-yd shovel less than optimum depth, estimate 75 cu yd $\times$ 2/3 = 50 cu yd)

Truck time: To maneuver and load = 8.33/0.83 + 0.5 = 10.5 min

               Round trip   = 7.0

               Delays and dump time   = 2.5

               Total trip time   = 20.0 min

Each truck can then make 60/20 or 3 trips per hour

Backhoe can load each truck in 8.33/0.833 = 10 min. Since it takes an extra half minute to get into position for loading, the number of trucks that can be loaded and moved away per hour = 60 / 10.5 = 5.71. And since each truck can make 3 trips per hour, the number of trucks required to keep the backhoe busy will be 5.71/3 = 1.9 trucks, which is close enough to 2 trucks in view of many variables in the computations. (Assume that there is enough extra work around project to keep backhoe and trucks busy for at least a week.)

4. Rental of backhoe based on a weekly rental rate is $700/40 = $17.50 per hour (see Appendix A).
Operating cost = $0.017 per brake hp × 125 hp (assumed) = $2.12
Rental plus operating cost per hour = $17.50 + $2.12 = $19.62

Rental cost of 10-cu-yd capacity diesel dump truck based on a weekly rental rate is $490/40 = $12.25 plus $0.34 for tire wear = $12.59. Operating cost = $0.017 per brake hp × 185 hp (assumed) = $3.14. Rental plus operating cost per hour per truck will be $12.59 + $3.14 = $15.73

5. Hourly rate for shovel operator $8.20
Hourly rate for each truck driver   6.20
Hourly rate for laborer   6.05
Hourly rate for foreman   8.55

6. Cost per cu yd of excavation then becomes:

   a.   Labor cost per cu yd

$$\frac{\text{operators wages/hr}}{\text{cu yd moved/hr}} = \frac{\$8.20 + (2 \times \$6.20) + \$6.05 + \$8.55}{2 \text{ trucks} \times 3 \text{ trips/hr} \times 8.33 \text{ cu yd}} = \$0.70$$

   b.   Equipment plus operating cost per cu yd

$$\frac{\text{total rental plus operating cost 1 hr}}{\text{cu yd moved/hr}} = \frac{\$19.62 + (2 \times \$15.73)}{2 \times 3 \times 8.33} = \$1.02$$

The solution for Examples 1 and 2 will now be shown on Quantity, Summary, and Direct Cost sheets. They will be treated as one job with overhead and profit added.

Estimate by _____ Date _____ Ckd. by _____ Date _____

Item	Identity	Location	Quantity Computations	Total	Units
	Classroom bldg.		Ex. 3-1		
300	Clearing and grubbing		About 20 trees per acre (8" to 12" dia)	3	acres
310	Gen. exc. Gen. exc.	Outside bldg. At bldg.	(200 x 160 x 2) /27 =	4,000 2,370	cu yd cu yd
	Basement		Ex. 3-2		
320	Gen. exc.	Basement	(200 x 160 x 8) /27 =	9,480	cu yd
330	Struc. exc.	Col. footings	[18(12 x 12 x 6) + 21(9 x 9 x 5) + 8(7 x 7 x 5)] /27 =	963	cu yd

SUMMARIES & UNIT COSTS—Example 3-1 and 3-2

Estimate by _____ Date _____ Ckd. by _____ Date _____

Item No.	Identity & Cost Source	Computation of Unit Costs	Total	Units
300	Clearing and grubbing	The 115-hp bulldozer, used to help load scraper, will be used with a blade to clear the trees. Equip. cost per hour is $15.74 Operator per hour is $ 8.20 Assuming 1/3 of foreman's time, the cost per hour is $8.55/3 = $2.85. Then labor cost per acre will be [($8.20 + $2.85) / ($15.74 + $8.20 + $2.85)] x $300 = $124 Equipment cost per acre will be [$15.74/ ($15.74 + $8.20 + $2.85)] x $300 = $176	3	acres
310	Gen. exc. See step 6, Ex. 3-1	4,000 + 2,370 = Cost per cu yd (see computation sheets): Labor $0.25 Equipment $0.34	6,370	cu yd
320	Gen. exc. See step 6, Ex. 3-2, Part 1.	 Cost per cu yd (see computation sheets): Labor $0.33 Equipment $0.71	9,480	cu yd
330	Struc. exc. See step 6, Ex. 3-2, Part 2.	 Cost per cu yd (see computation sheets): Labor $0.70 Equipment $1.02	963	cu yd

DIRECT COSTS—Example 3-1 and 3-2

Estimate by _____ Date _____ Ckd. by _____ Date _____

| Item No. | Identity & Location | Quantity | | Unit Cost Each | | | Total Cost Each | | | Total Cost |
		No.	Unit	Equip.	Mat'l.	Labor	Equip.	Mat'l.	Labor	
300	Clearing and grubbing	3	acres	$176.		$124.	$ 528		$ 373	$ 901
310	Gen. exc.	6,370	cu yd	0.34		0.25	2,166		1,593	3,759
320	Gen. exc.	9,480	cu yd	0.71		0.33	6,731		3,128	9,859
330	Struc. exc.	963	cu yd	1.02		0.70	982		674	1,656
	Direct costs						$ 10,407		$ 5,768	$ 16,175

OVERHEAD & PROFIT—Example 3-1 and 3-2

Estimate by _____ Date _____ Ckd. by _____ Date _____

Item No.	Class of Expense	Computations of Overhead Expense	Total Cost
1600	Gen. overhead (% direct cost)	2% of $16,175	$ 324
	Job overhead	Assume 30 days for job	
1700	Int. on operating capital	10% x $16,175/2 x 0.09 (int.)/12	5
1701	Superintendent's salary		
1702	Supt. pickup truck - rental	(charge 1/2 to excavation)	50
	do. operating cost		50
1703	Job trucks - rental		
	do. operating cost		
	do. wages - drivers		
1704	Lifting equipment		
	do. operating cost		
	do. wages - operators		
1705	Job office - rental		40
	do. salaries		
	do. supplies		25
1706	Utilities & connections		50
1707	Social Security	5.85% of $5768 x 90%*	304
1708	Workmen's Compensation	4% of $5768	231
1709	Pub. Lia. & Prop. Damage	0.2% of $5768 *	12
1710	Fed. & State Unemp. Ins.	3.2% x 45% of $5768	83
1711	Patents & royalties		
1712	Barricades		87
1713	Temporary toilets		50
1714	Cut and patch for trades		
1715	Permits		
1716	Protection adjacent prop.		
1717	Final cleanup		100
	Subtotal of computed overhead		$ 1,283
1718	Contingencies (% of computed overhead) 10% of $1,283		128
	Total overhead		$ 1,411
	Direct cost from Direct Cost sheet		$ 16,175
	Subtotal (Total overhead plus Direct Cost)		$ 17,586
1800	Profit 12% x $17,586		$ 2,110
	Total cost		$ 19,696
	Performance bond $19,696 x 1/1000 x $10		$ 197
	Total amount of bid		$ 19,893
	* Assume average wage of $7.00 per hour		

We will assume that this work is to be done by a subcontractor who maintains a storage yard for small equipment. He has an office at the yard with a secretary who keeps his books. He will require no storage facilities at the job site and will use the public services provided by the general contractor, who will also be responsible for permits. However, he will be required to show that he carries insurance covering Public Liability and Property Damage and Workmen's Compensation. And of course the payroll taxes consisting of Social Security and Federal and State Unemployment will be collected from him. He will also have to allow for the fact that the weather may become stormy or wet and cause him to lose rental time on the equipment. Then there will be labor losses since union rules require that the men be paid for several hours' work if they report even if it develops that no work can be done; and if they only work a short time, they must be paid for several hours or perhaps a half day. And to make matters worse this piecemeal work is unproductive.

The profit in this case should be 15%. But assuming there is a great deal of competition we will only figure the profit here is 12%.

HOMEWORK EXAMPLE 3-1

Using the following sketch plan (Fig. 3-3) compute the bid price for the area excavation and for the excavation within the building limits and also for the clearing and grubbing. Assume 4 acres of clearing with approximately 30 trees per acre of 6-inch to 12-inch diameters. The building will be the slab on grade type with a 5-inch concrete slab on a 5-inch gravel fill. The soil will be hard clay with a swell of 20%.

The area excavation and the excavation inside the building limits to the bottom of the gravel fill will be considered general excavation and will be used to bring the area to finished contour elevations, being worked into place by bulldozer as the excavation proceeds. The general excavation will be computed from the contour lines. The sections to compute excavation should in no case be taken at intervals greater than 50 feet. Let these sections continue through the building with the finished grade within the building limits being a little below the bottom of the gravel fill, say at elevation 99'–4". (See note bottom of page 48.)

This is necessary because there is a fill required at two of the corners of the building as shown on the plan. The slab foundation should not be partly on fill and partly on cut since this would give uneven bearing and cause the floor to crack along the line where the fill joins the cut. This can be prevented by having a little cut over the entire area which can be compacted uniformly.

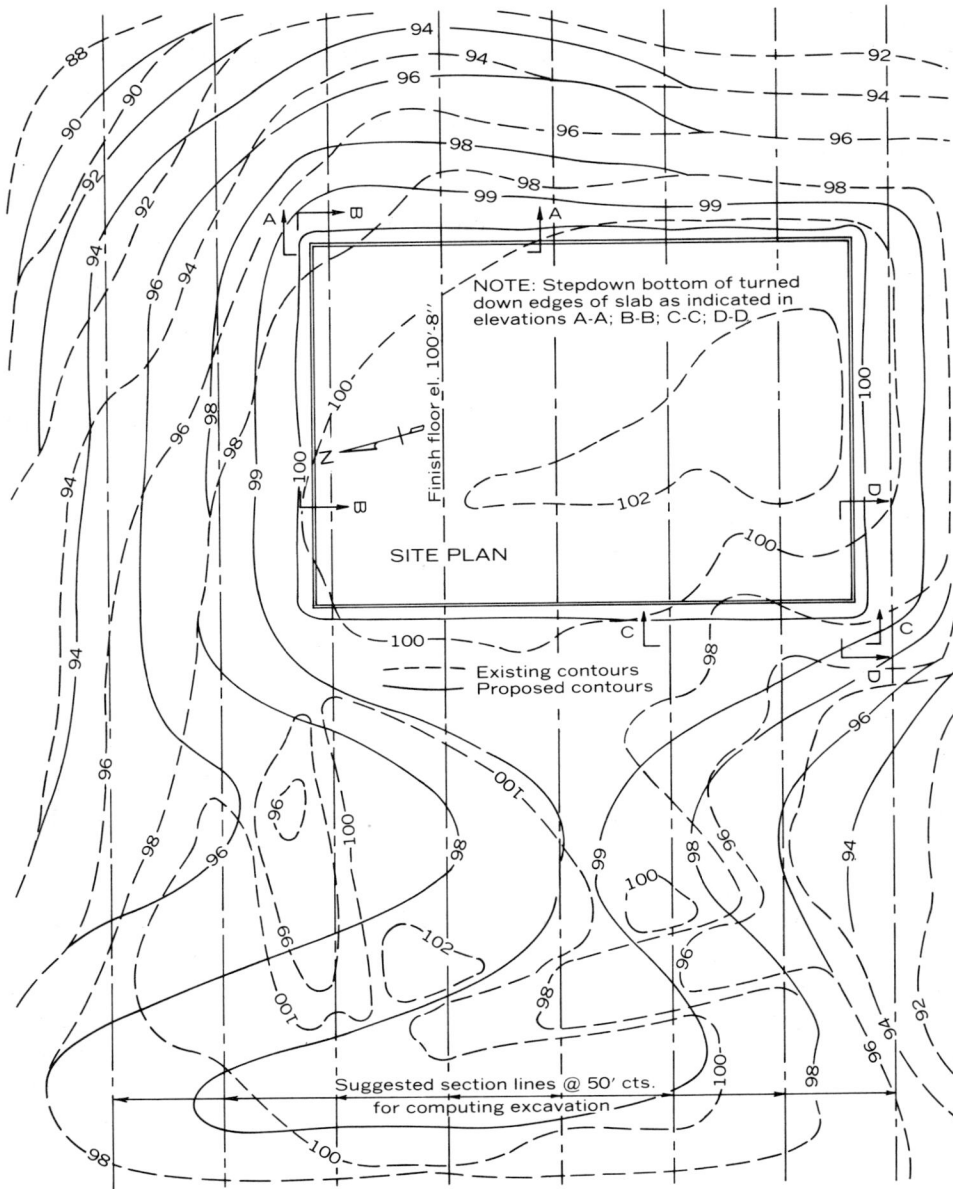

NOTE: Stepdown bottom of turned down edges of slab as indicated in elevations A-A; B-B; C-C; D-D.

Finish floor el. 100'-8"

SITE PLAN

- - - - Existing contours
———— Proposed contours

Suggested section lines @ 50' cts. for computing excavation

FIGURE 3-3
Sketch Plan for Homework Example 3-1

A. Site Plan

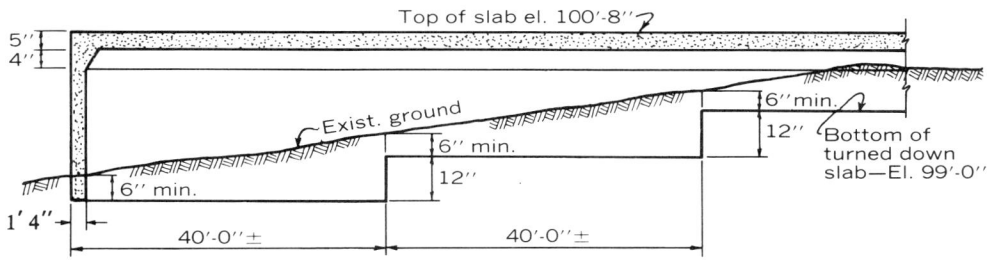

ELEVATION A-A

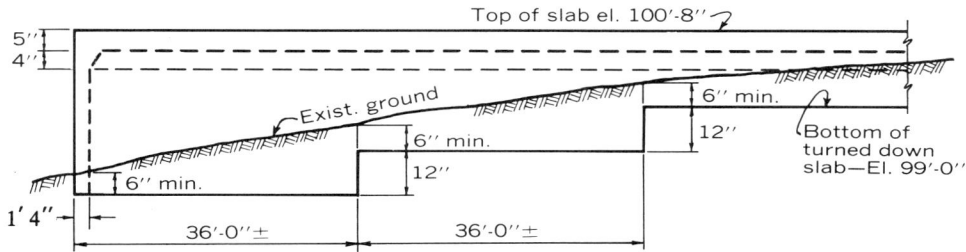

ELEVATION B-B

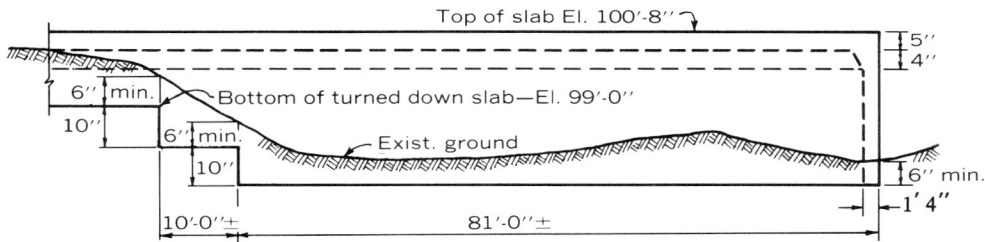

ELEVATION C-C

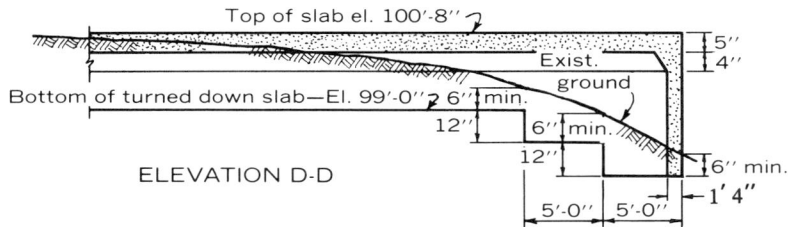

ELEVATION D-D

FIGURE 3-3 (continued)

B.   Elevations

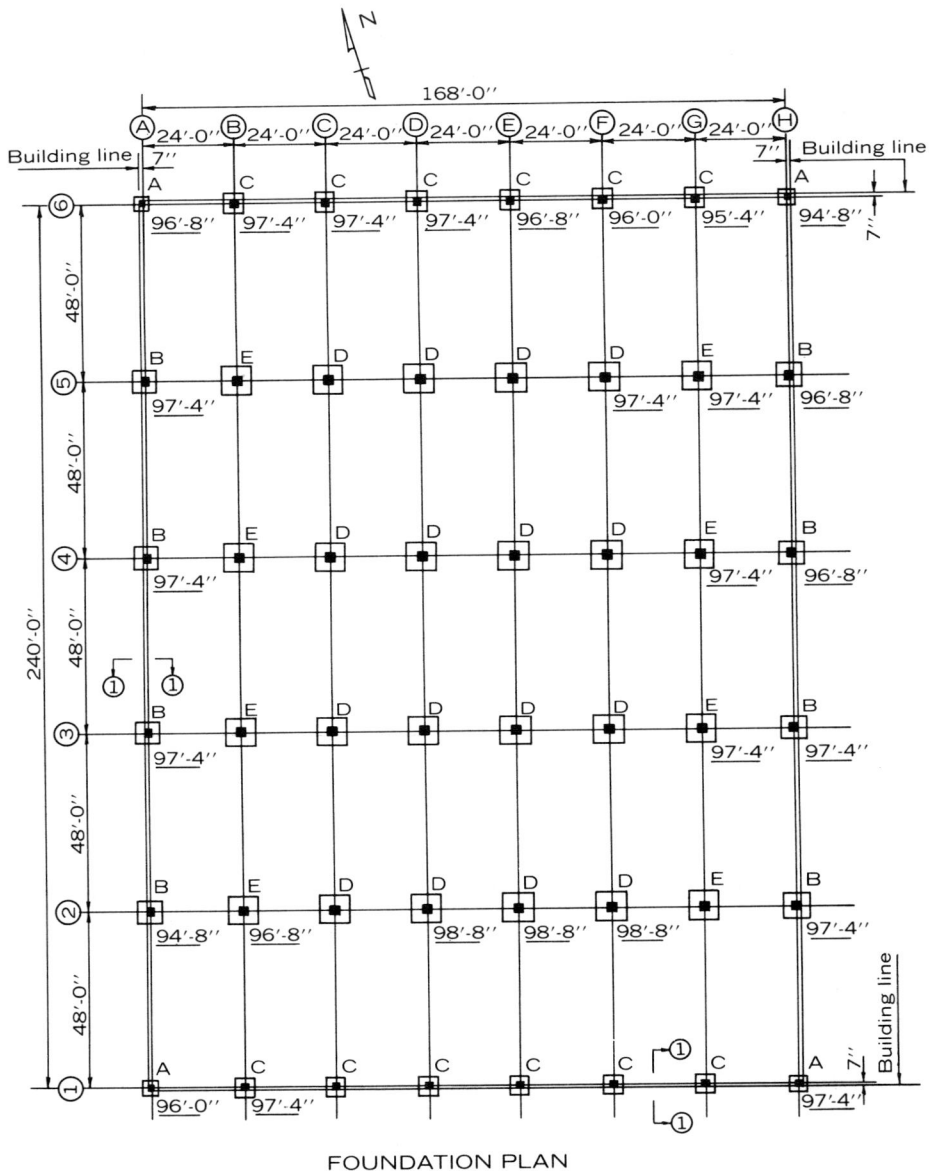

FOUNDATION PLAN

FIGURE 3-3 (continued)

C.   Foundation Plan

46   *Chapter Three*

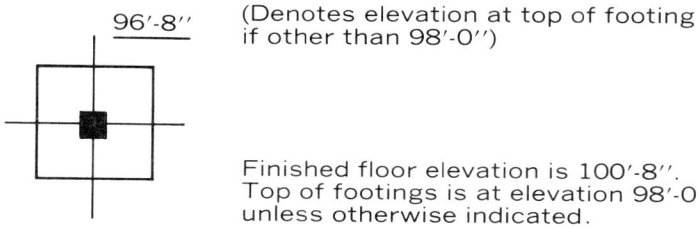

96'-8'' (Denotes elevation at top of footing if other than 98'-0'')

Finished floor elevation is 100'-8''.
Top of footings is at elevation 98'-0''
unless otherwise indicated.

D. Top of Footing Elevations

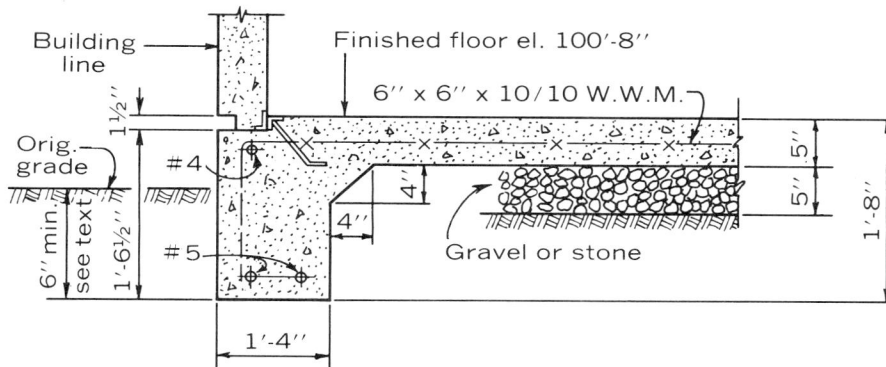

Building line

Finished floor el. 100'-8''

6'' x 6'' x 10/10 W.W.M.

Orig. grade

#4

#5

Gravel or stone

6'' min see text

1'-6½''

1'-4''

4''

4''

1½''

5'' 5''

5'' 5''

1'-8''

SECTION 1-1

E. Section 1-1 Through Turned-Down Slab

NOTE: Footings were designed for a min.
bearing value of 4000 lbs. per sq. ft.

FOOTING SCHEDULE					
Mark	A	B	C	D	E
Size	3'-6''x3'-6''	4'-0''x4'-0''	4'-6''x4'-6''	5'-0''x5'-0''	5'-6''x5'-6''
Depth	1'-0''	1'-0''	1'-3''	1'-3''	1'-3''
Reinf.	9 #4 E.W.	10 #4 E.W.	12 #4 E.W.	9 #5 E.W.	11 #5 E.W.

(See foundation plan)

F. Footing Schedule

Refer to Ex. 3-1 for guidance in estimating the cost of the area excavation.

The structure excavation will be composed of the volumes formed between the bottoms of the footings and the turned-down portion of the slabs and elevation 99'-4".

Refer to Ex. 3-2, part 2, in estimating the cost of the structure excavation.

It should be noted that the bottom of the turned-down edges of the slab must be kept at a minimum of 6 inches below the original ground to insure a solid bearing below top soil or to a depth necessary to obtain the design bearing value, which in this case will be 4,000 lb per square foot. (Bottom of footings must also be below the frost line.) See note on plan and sections.

In choosing the backhoe size to be used several things should be kept in mind:

1. The footings are quite small and the turned-down edge of the slab is only 16 inches wide. A small size of backhoe should be used—such as one with a 3/8-cu-yd bucket or perhaps even smaller.

2. The optimum digging depth at which these quantities are possible is 6'-0', while the depth of these footings is an average of less than three feet. This fact would reduce the capacity shown in Table 3-2 from 50 to perhaps 40 cu yd per hour in hard clay.

3. The table was computed on the basis of continuous, 60 minutes' per hour digging. Since there are 48 separate footings each requiring a separate setup and since there are also about 760 linear feet of turned-down slab, also necessitating quite a few setups, then only about 45 minutes per hour can be considered as productive time. This would reduce the output to 40 x 0.75 = 30. This table was made for power shovels; therefore, the result must be multiplied by 2/3 to convert to backhoe capacity, 2/3 x 30 = 20 cu yd per hour. In excavating for the turned-down edges of the slab this capacity would be much less because the 3/8-cubic-yard-capacity bucket must be removed and replaced with a smaller bucket, only 16 in. wide, whose capacity will perhaps be 10 cu yd per hour.

The excavation for the footings and the turned-down slab will be considered as structure excavation and will be stockpiled to be later used as backfill.

Note: For a method of obtaining excavation quantities from the contour lines on Fig. 3-3 see page 27. Plot sections at the eight indicated section lines in this manner: Lay a strip of paper along section line and mark off points where contours cross the line and then transfer to cross section paper. Read distances by scale made to match figure by using the 50-foot intervals between section lines.

CHAPTER FOUR

Concrete

There are six principal costs which must be considered in computing the cost of concrete in place:

1. Forms
2. Reinforcing steel
3. Concrete
4. Placing concrete
5. Finishing
6. Curing

These items will be taken up in order.

FORMS

The formwork has more influence on the cost of concrete in place than any other factor. If the beam and column sizes are repeated enough, then the forms can be used over again from three to five times and even more. Every time the forms are reused the cost of formwork per cubic yard of concrete is reduced. The forms only have to be assembled into panels one time regardless of the number of uses; the only work that has to be repeated for each additional use is the erection, stripping, and cleaning.

To illustrate the effect on the cost due to reuse of the forms let us consider the beams in Fig. 4-4 for Example 4-6 (shown in section c-c, p. 66, indicated as item 430). Refer to Summary sheets for Example 4-6 where the computations are shown for three uses of the forms. We will compare the price per cubic yard of concrete in place for one, three, and five uses of the forms assuming no salvage value of the formwork in either case.

Use costs for materials, labor, and equipment shown on the Summary sheet and adjust for the number of times forms are to be used.

1. For one use of forms:

Materials cost—	=	$ 69.68
add for single use of forms = $10.96 × 2	=	21.92
Labor cost—	=	51.00
add for assembly = $8.89 × 2	=	17.78
Equipment cost	=	2.79
Total cost per cu yd		$163.17

2. For three uses of forms:

Materials cost	=	$ 69.68
Labor cost	=	51.00
Equipment cost	=	2.79
Total cost per cu yd	=	$123.47

3. For five uses of forms:

Materials cost—   = $ 69.68

subtract for 5 uses of forms = $\dfrac{\$10.96 \times 2}{5}$ = (—) 4.38

Labor cost—   = 51.00

subtract for assembly = $\dfrac{\$8.89 \times 2}{5}$ = (–) 3.56

Equipment cost   = 2.79

$115.53

Thus we see that three uses of the forms will save $39.70 per cu yd and five uses will save $47.64.

Formwork must be capable of sustaining erection loads consisting of the weight and side pressure of the green concrete, equipment, personnel, etc.; it must be rigid enough to maintain its shape without undue deflection; and it must be economical in terms of the final cost.

Forms should be designed by conventional engineering methods to resist the forces resulting from the known loads. If designed by experience or guesswork the forms might be too light and subject to an expensive failure. On the other hand the forms could be stronger than necessary and cost too much.

When the forms are to be reused many times, then that part of the formwork in contact with the concrete must be of a good grade of material which will have a high first cost but will still be economical due to more than one use. If the finished surface of the concrete must be smooth and free of blemishes, then it would be feasible to use a very high grade of material to minimize expensive finishing.

The successful bidder will usually be responsible for the design and preparation of the falsework plans; however, he can estimate the necessary fbm of lumber without having to do all of this work. Tables can be made showing the approximate amount of lumber and labor hours required for the forms, braces, supporting beams, and shores that make up the formwork.

Such tables are usually based upon the amount of contact area between the concrete and the forms and show the total fbm of formwork and labor hours required per 100 square feet of contact area. Sample tables of this type are shown in this text. It should be realized that all contractors do not build their forms exactly alike and therefore the fbm per 100 sq ft of contact area will vary to some degree. Because of this these tables should be altered to fit the work of the contractor for whom the estimate is being prepared.

In general the contact areas should be measured as follows:

For footings—sides of footing (and bottom if required) in sq ft

For floors—total area of bottom of slab in sq ft

For columns—perimeter of columns times the height between floors in sq ft

For walls—areas of each side of wall in sq ft

For beams and girders—areas of sides and bottoms between supports in sq ft

For molding, lintels, etc.—measured in linear feet

Hardware is another item that must be considered when estimating form material. There are special two-headed nails used on forms. The nail is driven to its first head. This leaves a part of the nail with the second head protruding from the wood so that a wrecking bar or claw hammer can engage the head when stripping the forms. Figures 4-1, 4-2, and 4-3 show "contact surfaces" and some of the hardware types, including: nuts, bolts, washers, nails, wall ties, etc.

Tables 4-1 and 4-2 show the approximate amount of materials and labor required per 100 square feet of contact area for several different types of members. The estimator should prepare such tables as these from the work experience of his own organization.

The price of form hardware will vary in the range of $0.25 to $0.35 per pound. The cost of form lumber will vary from $90 to $250 per 1,000 fbm depending on the grade and species of the lumber and the locality of the project. The cost of ordinary 5/8" plywood for formwork will be between $90 and $250 per 1,000 sq ft and for 3/4" plywood will be between $125 and $300 per 1,000 sq ft. Special type plywood of marine grade will cost in the neighborhood of $50 more per 1,000 sq ft. These prices will be followed in developing the examples shown in the text, but the student should use local prices in doing the homework examples.

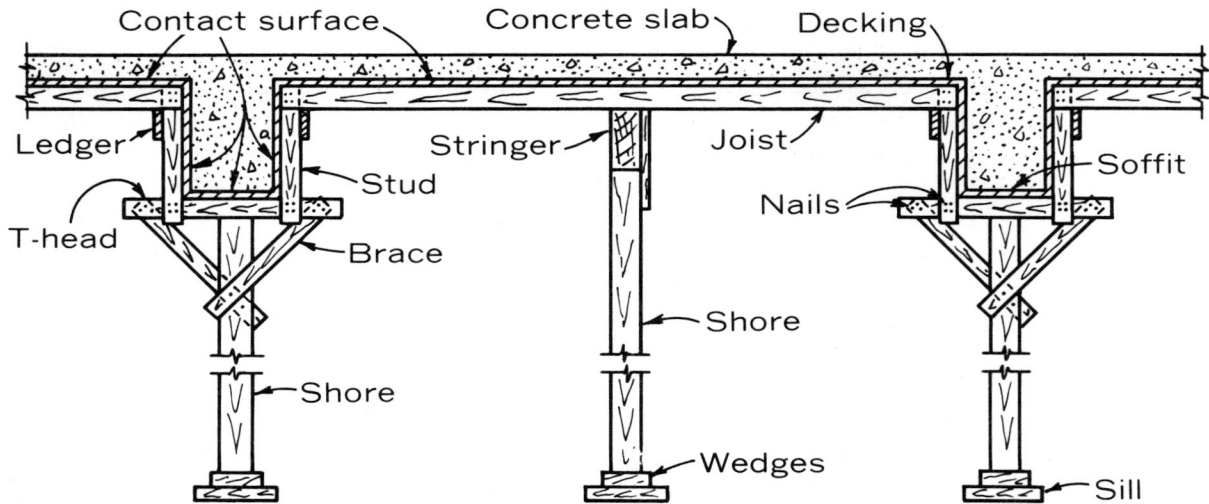

FIGURE 4-1
Wood Forms for Beam and Slab Floor

Forms have to be oiled in order to keep the concrete from sticking to them. A gallon of oil will cover about 400 sq ft and cost approximately $1.25 per gallon. It will probably be applied by a carpenter helper during the course of his regular work on forms and so it is not usually necessary to allow any extra labor for applying the oil.

In our examples we will assume two to four uses for lumber formwork with no salvage value, and three to five or more for plywood formwork according to type and grade.

In this text we will use wooden forms in our examples. But there are many different types of metal forms and special hardware available. In the case of pan and joist construction the cost of erecting, stripping, and cleaning the forms may not differ too much from wooden forms. However, if the metal forms can be reused enough times there will be a saving in cost over wood and a saving in job time because the metal forms do not have to be fabricated.

Other types of metal forms, most of which are patented, are: prefabricated panels for wall forms, adjustable metal shores, adjustable column clamps, and floor forms of many types which are usually left in place. As a rule there is not too much difference in their use as far as overall cost is concerned. Each contractor has his own preference as to which he will use and which he believes is the cheapest.

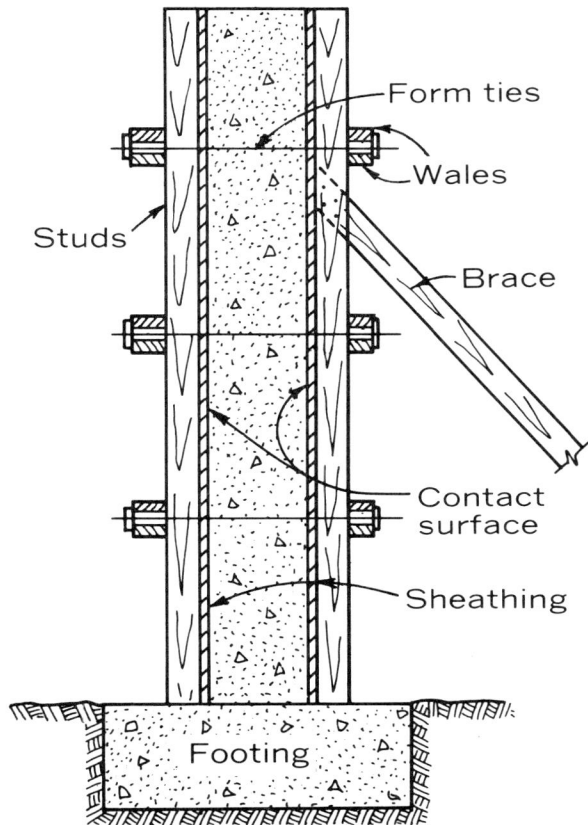

FIGURE 4-2
Wood Forms for Concrete Wall

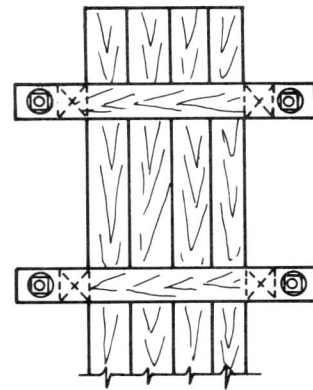

ELEVATION

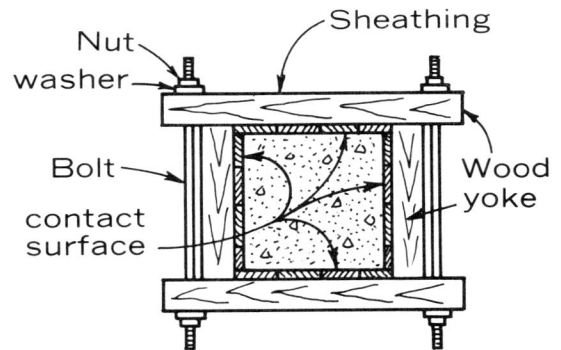

FIGURE 4-3
Wood Column Forms

Formwork is most efficiently and economically fabricated and erected by a gang composed of carpenters, helpers, and laborers operating under the direction of a carpenter foreman. One good division of workmen in a gang would be:

1	Foreman	@	$8.30	=	$ 8.30
3	Carpenters	@	7.85	=	23.55
3	Apprentices	@	6.50	=	19.50
3	Laborers	@	6.05	=	18.15
	Total for gang				$69.50

TABLE 4-1

Approximate Quantities of Formwork Required
per 100 Square Feet of Contact Surface for Wooden
Forms

Location	Lumber (fbm)	Hardware (lb)
Footings	190–340	5–10
Walls	190–260	5– 8
Columns	180–300	6–10
Floors	175–275	6– 8
Roofs	190–280	6–10
Stairs	320–600	9–15
Beams and girders	280–680	9–14
Moldings, cornices, sills, lintels, etc.*	200–700	6–14

* These are quantities in fbm per 100 linear feet.

TABLE 4-2

Approximate Labor Hours Required per 100 Square
Feet of Contact Surface for Wooden Forms

Location	Assemble	Erect	Strip and clean
Footings	3– 6	2–4	2–4
Walls	5– 8	3–5	2–4
Columns	4– 7	2–5	2–5
Floors	3– 7	2–4	2–4
Roofs	3– 8	2–4	2–4
Stairs	6–11	4–7	3–4
Beams and girders	6–10	3–4	2–4
Moldings, cornices, sills, lintels, etc.*	5– 9	3–8	3–5

*These are labor hours required per 100 linear feet.

To get the average hourly wage add the hourly wages of carpenter, helper, laborer, and foreman and divide by 10. The average hourly wage in this formwork gang would be $69.50/10 = $6.95.

On a project where much falsework is required a carpenter shop will be set up with power tools to fabricate the forms. This cost will usually vary from $1.00 to $3.00 per 100 sq ft of contact area, depending upon the amount of equipment being used. This power equipment will reduce labor hours for fabricating wooden formwork.

REINFORCING STEEL

There are two types of reinforcing steel: steel bars and welded wire fabric. The welded wire fabric usually comes in rolls but can be obtained in flat sheets. It is made of cold-drawn steel wire welded at intersection of the transverse and longitudinal wires. The area required will be the area of floor being reinforced plus an allowance for laps of 5 to 10%. It is usually priced per roll. It is relatively simple to place. A good workman should place 400 sq ft per hour on large areas that are not excessively divided, but on small areas where much cutting or fitting is necessary he will only be able to do about half as much.

Reinforcing steel bars are in the vast majority of cases fabricated to exact lengths and bent as required by bar fabricating companies, who are expert at this type of work and have all the necessary laborsaving machinery. It is usually possible to get a direct quotation from one of these shops for the reinforcing steel fabricated and delivered to the job. However, the variables governing the cost of the reinforcement will be discussed because the estimator should be able to check the pound price of the different bar sizes given to him by the fabricating company.

Table 4-3 shows the size and weight of reinforcing bars. It will be noticed that the bar numbers, by which the bars are designated on the plans, show the bar diameters in eighths: for instance, a No. 4 bar means a bar having a diameter of 4/8" or 1/2" and a No. 8 is a bar of 8/8" or 1"-diameter. Bars #9, 10, and 11 are approximately 1 1/8", 1 1/4", and 1 3/8" in diameter and replace in weights and areas the former square bars of those dimensions which are no longer made.

The base price of steel reinforcing bars as of August 5, 1971, varies from $8.85 to $9.65 per 100 lb at the mill; to this must be added extras as shown in Example 4-1.

TABLE 4-3
Physical Properties of Reinforcing Bars

Bar number	Diameter (in.)	Area (sq in.)	Weight (lb per ft)
2*	1/4	0.049	0.167
3	3/8	0.110	0.376
4	1/2	0.196	0.668
5	5/8	0.307	1.043
6	3/4	0.442	1.502
7	7/8	0.601	2.044
8	1	0.785	2.670
9	1 1/8	1.00	3.400
10	1 1/4	1.27	4.303
11	1 3/8	1.56	5.313
14S	2 3/4	2.25	7.65
18S	2 1/2	4.00	13.60

*No. 2 bars are plain round bars; all of the remainder are deformed bars.

EXAMPLE 4-1

APPROXIMATE COST OF 15 TONS OF #6 AND LARGER BARS IN PLACE

Item	Cost per 100#	
Base price at Richmond as of August 5, 1971	$9.20	
Freight to Raleigh, N.C.	0.50	
Quantity extra	1.25	
Size extra	0.50	
Detailing	1.15	
Bending	1.50	
Delivery to the job	0.20	
Reinforcement supplier's overhead and profit-12%	1.72	
Chairs, spacers, ties, etc.	+0.50	= $16.52
Erection by rodbusters $91.15 per ton		= 4.56
Total cost in place		$21.08

Since we are keeping material and labor separate we would in this case carry the $16.52 figure forward to the final estimate sheet as material and the $4.56 figure as a labor item. Of course there is labor in the $16.52, which is the cost per 100 lb of the fabricated steel fob job site, but it is a figure quoted to the contractor by the reinforcement supplier and therefore a material as far as the contractor's costs are concerned because the supplier has included in this price the payroll taxes and insurance costs incident to the work.

Tables 4-4 through 4-9 will supply the data which were used to compute the reinforcing bar prices.

In order to hold reinforcing bars firmly in place while concrete is being cast and vibrated it is necessary to have bar supports consisting of chairs, spacers, ties, etc., to which the bars may be secured. Approximate cost will be $0.50 to $1.50 per 100 lb.

TABLE 4-4
Base Prices per 100 Lb (Before Extras Are Added)
for Deformed Concrete Reinforcing Bars fob Plant

Albany, N.Y.	$9.30
Baltimore, Md.	9.00
Boston, Mass.	9.65
Clearing, Ill.	9.05
Cleveland, Ohio	9.05
Detroit, Mich.	9.25
Elizabeth, N.J.	9.30
Houston, Tex.	9.00
Jacksonville, Fla.	9.20
Lackawanna, N.Y.	8.90
Los Angeles, Calif.	8.85
Miami, Fla.	9.15
Minneapolis, Minn.	9.25
New Haven, Conn.	9.40
Philadelphia, Pa.	9.15
Richmond, Va.	9.20
Seattle, Wash.	8.85
South San Francisco, Calif.	8.85
Steelton, Pa.	8.85
Washington, D.C.	9.15

TABLE 4-5
Quantity Extras per 100 Lb of Reinforcing Bars*

Under 20 tons	$1.25
20 to 49 tons	0.75
50 to 99 tons	0.50
100 to 300 tons	0.25
Over 300 tons	None

*Quantity extras apply on the total weight of an order (less spirals) regardless of the number of sizes involved. The price of column spiral bars is obtainable from the steel companies on request.

TABLE 4-6
Size Extras per 100 Lb of Reinforcing Bars

Size of bar	Extra
#2	$2.50
#3	1.40
#4	1.00
#5	0.70
#6 through #11	0.50
#14 through #18	0.50

TABLE 4-7
Cost of Detailing and Placement Plans per 100 Lb
of Reinforcing Bars

Under 50 tons	$1.15	Min. charge	$ 200
50 to 150 tons	0.85	Min. charge	1,150
150 to 500 tons	0.75	Min. charge	2,550
500 to 1,000 tons	0.60	Min. charge	7,500
1,000 tons and over	0.55	Min. charge	12,000

TABLE 4-8
Bending Extras per 100 Lb of Reinforcing Bars

Heavy bending (#4 bars and larger—bent at no more than 6 points)	$1.50
Light bending (#2 and #3 bars—bent at no more than 6 points)	2.25

TABLE 4-9
Approximate Cost of Placing Steel Reinf. Where Bars Are Tied in Place

	Hours/Ton	Average wage	Cost/Ton
For #5 bars and smaller			
Ironworkers	23	$8.64	$198.72
Rodbusters	18	8.44	151.92
For #6 bars and larger			
Ironworkers	18	8.64	155.52
Rodbusters	12	8.44	101.28

After fabrication the reinforcing bars are usually transported to the job site by truck. This cost will never be less than $0.15 per 100 lb and will usually be higher.

There is a considerable difference in the amount of reinforcement that an ironworker and a rodbuster can place because placing reinforcing bars is an art in itself in which the rodbuster excels just as the ironworker is more accomplished in setting structural steel. There is also a difference in the tonnage that will be handled and placed due to the weight of the bar, for the simple reason that the same number of #6 bars will weigh a lot more than #4 bars, which will require approximately the same time for handling and placing. Tables 4-9, 4-10, and 4-11 will demonstrate these variations.

If the steel does not have to be tied then these costs will be about 10% less in each case.

When welded wire fabric is to be placed in the floor slab for strength, to resist stresses due to changes in temperature, or for other purposes the approximate cost will be as shown in Table 4-10.

TABLE 4-10
Cost of Placing Welded Wire Fabric

Hours/100 sq ft	Average wage hour	Labor/100 sq ft
.5	$8.44	$4.22

In large areas where little cutting and fitting are required to install the wire fabric these figures can be reduced.

The average hourly wage used here was arrived at by using a gang of one foreman and 3 ironworkers:

$$\$8.90 + (\$8.55 \times 3)/4 = \$8.64 \text{ per hour}$$

or a gang of one foreman and 3 rodbusters:

$$\$8.70 + (\$8.35 \times 3)/4 = \$8.44 \text{ per hour}$$

These prices are approximate and are to be used for instruction only.

TABLE 4-11
Spirals

Base cost fob cars or trucks at Fabricating Plant at Clearing, Ill., Philadelphia, Pa., and Richmond, Va:

Hot rolled, assembled	$19.00 per 100 lb
Cold drawn, assembled	20.50 per 100 lb

Quantity extras

under 2 1/2 tons	2.00 per 100 lb
2 1/2 to 9 tons	1.50 per 100 lb
10 to 14 tons	1.00 per 100 lb
15 to 20 tons	0.50 per 100 lb
over 20 tons	None

Size extras

under 30 inches	None
30 in. to under 64 in.	$1.00
64 in. to under 84 in.	2.00
84 in. to 96 in. max.	2.50

Table 4-11 shows the base cost and extras for spiral reinforcing used in columns. The spirals usually come from the mill already bent into the required shape. Although the spirals come prefabricated an allowance must be made for the crane to lift the reinforcing cages into placed and also for the labor for handling.

CONCRETE

Practically all of the concrete used in construction today is bought from ready-mix plants. These suppliers buy the sand, stone, and cement in such large quantities that they obtain the very lowest prices; they employ the best of laborsaving equipment and are in continual production. Because of these factors the ordinary contractor cannot begin to produce concrete as cheaply as he can buy it delivered from the ready-mix plant.

Many variables govern the price of ready-mix concrete such as: the cost of transportation when the job is farther from the plant than the freehaul distance; local and state sales costs and other applicable taxes. An overall average figure of $20 per cubic yard will be used in this text for 3,000 psi concrete. Concrete of 2,500 psi will cost about $0.50 less per cu yd and strengths higher than 3,000 psi will cost more. The estimator should obtain the exact cost for the area in which the work is being done.

Very large construction projects such as dams and very long concrete bridges and work in undeveloped areas are the only cases where ready-mix concrete is not now used.

Estimating Concrete Quantities

In taking concrete quantities from the plans it works best to have a definite system in order not to overlook any item. One very good order is to start with the foundations and continue on through to the roof in this order:

1. Foundation
   a. Footings
   b. Grade beam
   c. Turned-down slabs
2. Slabs
   a. Slab on grade
   b. Pan and joist
   c. Waffle slab
   d. Beam and slab
   e. Flat slabs

3. Drop panels
4. Beam and girders
5. Columns
6. Partitions
7. Exterior walls
8. Window sills
9. Lintels
10. Copings
11. Stairs
12. Canopies
13. Roof systems
14. Miscellaneous
15. Walks and drives
16. Curb and gutter

PLACING CONCRETE

The cost of placing concrete in the forms and properly vibrating it into position varies with many factors. The following analysis covers some of these costs.

Each of the following examples will be based upon the assumption that concrete will be delivered to the job and placed at the rate of 15 cu yd per hour.

EXAMPLE 4-2
COST OF PLACING CONCRETE ON UPPER FLOORS

A probable gang to handle 15 cu yd per hour on an upper floor could be:

1 man to handle crane bucket on the ground while it is being filled from ready-mix truck
1 man to guide crane bucket as it empties into concrete hopper on floor being cast
1 man opening hopper gate to fill buggies which will wheel concrete into place
5 men pushing these buggies
2 men helping to manhandle the buggies into correct position for dumping
5 men spreading the concrete into place and screeding the surface

4	utility men to shift runways, etc.		
1	man to operate vibrator		
20	concrete workers @ $6.15 per hour	=	$123.00
1	carpenter to take care of runways and forms @ $7.85 =		7.85
1	foreman @ $8.30	=	8.30
1	crane operator @ $8.20	=	8.20
			$147.35

Labor cost $147.35/15 cu yd = $9.82 per cu yd

In computing equipment cost use monthly rental basis since a crane will be required during the entire job to lift reinforcing bars, formwork panels, mechanical equipment, etc., to the upper floors.

Using a 20-ton diesel-powered crawler crane the rental cost per hour will be

$$\frac{\$1,803 \text{ per month}}{176 \text{ hours (in month)}} = \$10.24$$

Cost of fuel per hour = 2/3 × 150 hp × 0.055 gal × 0.30   = $ 1.65
Cost of oil and grease per hour = 55% × $1.65   = 0.91
$12.80

Equipment plus operating cost = $12.80/15 = $0.85 per cu yd

The unit prices in Ex. 4-2 will apply to all concrete poured above the ground floor to the last floor which can be reached by the crane. For buildings above this height the concrete will have to be carried up by a lift and the placing prices per cubic yard will change.

EXAMPLE 4-3
COST OF PLACING FLOOR SLAB ON GRADE

When the floor slab is at ground level it will require nine less men than for upper floors because the crane bucket can be emptied directly into the forms and buggies will not be required. In this case the unit costs per cubic yard will become:

11	concrete workers @	$6.15 per hour	=	$67.65
1	carpenter @	7.85	=	7.85
1	crane operator @	8.20	=	8.20
1	foreman @	8.30	=	8.30
				$92.00

Labor cost = $92.00/15 cu yd = $6.13 per cu yd
Equipment cost will be the same as in Ex. 4-2 = $0.85 per cu yd

The unit prices in Ex. 4-3 will also apply to concrete poured in foundation walls and grade beams.

EXAMPLE 4-4
COST OF PLACING CONCRETE IN LARGE FOOTINGS

If the concrete is to be placed in large footings where the delivery truck can back up to the footings and discharge directly through a chute, then the labor crew will not have to be nearly so large as for pouring on upper floors. In this case, a foreman, a carpenter, and seven concrete workers should be ample to pour 15 cu yd of concrete per hour and the costs per cubic yard can be determined as follows:

7	concrete workers @	$6.15 per hour	=	$43.05
1	carpenter @	7.85	=	7.85
1	foreman @	8.30	=	8.30
				$59.20

Labor cost per cu yd = $59.20/15 cu yd = $3.95

EXAMPLE 4-5
COST OF PLACING CONCRETE IN BRIDGE PIERS

For placing concrete in bridge piers the labor force will be the same except for the addition of a crane operator @ $8.20 per hour.
Labor cost will be ($59.20 + $8.20)/15 = $4.49 per cu yd
Equipment cost will be same as in Ex. 4-2 = $0.85 per cu yd

FINISHING CONCRETE

Finishing concrete is another cost factor that must be considered. Table 4-12 shows approximate man hours required for certain types of finishing.

TABLE 4-12
Cost of Finishing Concrete Surfaces

Type of work	Labor hours per 100 sq ft
Troweling floors, walls, sidewalks, etc.	2– 4
Rubbing floors, walls, etc., with carborundum	4–10
Grinding with machine	4–12
Sand blasting	3– 7

A finishing team could be a finisher at $7.70 per hour and a helper at $6.15 making an average of $6.93 per hour.

CURING CONCRETE

The cost of curing concrete depends to a large extent upon the weather. In cold weather the cost can equal that of placing the concrete and in addition to this it may even be necessary, if the weather is very cold, to heat both the aggregate and the water before mixing. This cost of heat may vary from $1 to $6 per cu yd or more.

Some of the methods by which concrete can be cured during hot weather are: spraying on a waterproof membrane; using a cover of a material such as burlap which is kept wet during the curing period; applying a cover of straw or earth; and also by leaving the wooden forms in place during the curing period. Any one of these methods is much cheaper than the cold weather curing during which the concrete must be housed in a protective enclosure which is warmed by artificial heat.

The cost of curing during hot weather will likely be in the range of $0.10 to $0.50 per cu yd. The cost of curing during cold weather can vary from $1.00 to $6.00 or more per cu yd.

EXAMPLE 4-6

Take off the framing material and compute unit costs for four stories of a 20' × 41' reinforced concrete building (stop at the finished fourth floor elevation). See Fig. 4-4. The finished first-floor elevation will be 3'-0" above the top of the column footings. The first floor will be slab on grade with outside edges turned down 2 feet from top of slab. The turned-down portion will be 1'-0" wide.

The building has columns at each corner with two beams the long way between columns and five beams the short way spaced at 10'-0" cts (cuts). Each floor is framed alike with the same size beams, columns, and slabs.

MATERIAL

Since we are using the unit price method we will take off the cubic yards of the concrete only and record on the Quantity sheets. Then these quantities will be carried to the Summary sheets where the concrete in each type of member will be totaled.

On the Summary sheets we will compute the cost per cubic yard for all of the materials, labor, and equipment required for one cubic yard of each type of concrete in place. Then these quantities and unit prices will be carried to the Direct Cost sheets and extended.

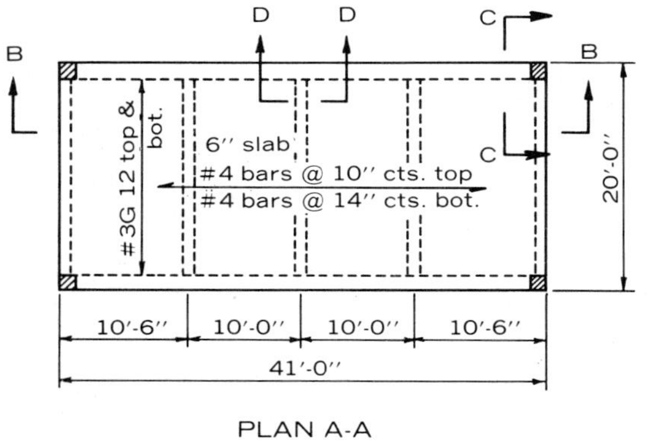

PLAN A-A

A. Plan A-A

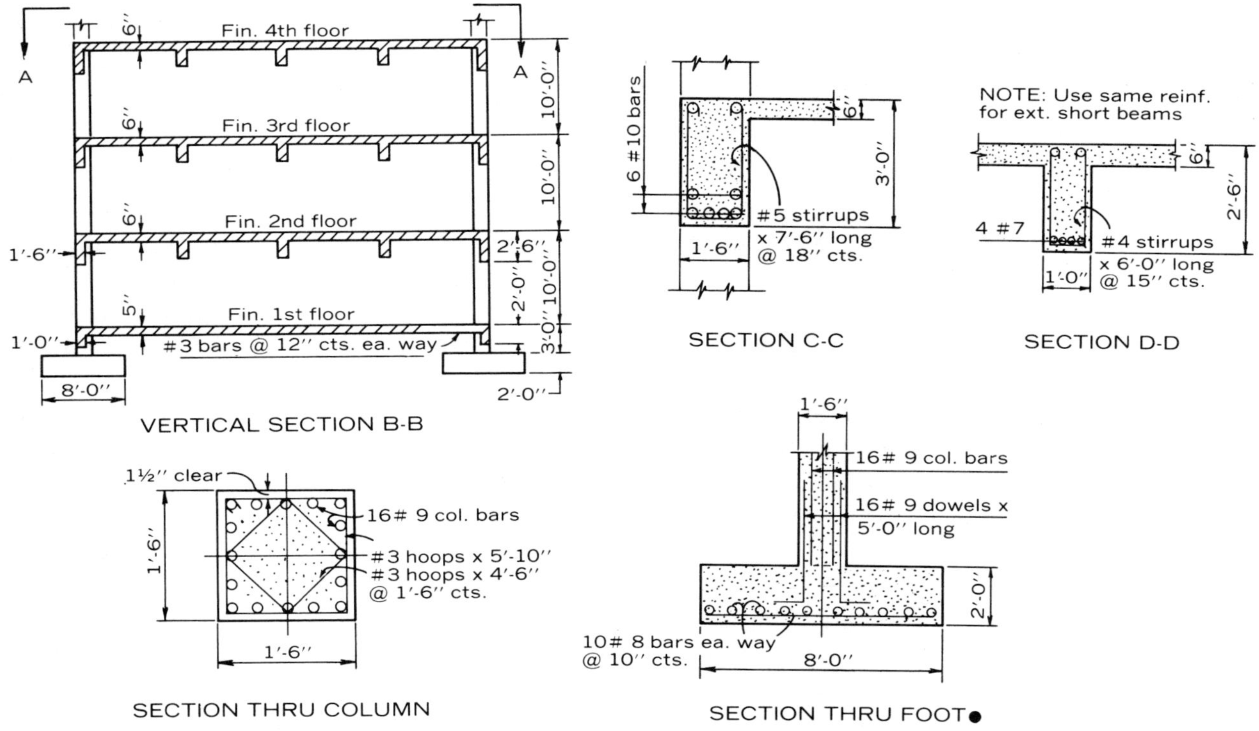

VERTICAL SECTION B-B

SECTION C-C

SECTION D-D

SECTION THRU COLUMN

SECTION THRU FOOTING

B. Sections (Sections B-B, C-C, D-D, section through column, section through footing)

FIGURE 4-4
Plan and Sections for Example 4-6

Before attempting to estimate the cubic-yard cost of the several concrete members let us review the data which we have established for the purpose of solving the six cost items of concrete construction.

1. Forms

   Table 4-1 estimates the required fbm of formwork per 100 sq ft of contact area.

   Table 4-2 gives the approximate labor hours required per 100 sq ft of contact area of the forms.

2. Reinforcing Steel

   Table 4-3 lists weights per linear foot of reinforcing bars.

   Example 4-1 illustrates method of computing pound price of reinforcing bars delivered to job site by use of Tables 4-4 to 4-9.

   Table 4-4 lists base prices per 100 lb of reinforcing bars at various plants.

   Table 4-5 gives the quantity extras per 100 lb of reinforcing bars.

   Table 4-6 gives size extras per 100 lb of reinforcing bars.

   Table 4-7 shows approximate cost of detailing per 100 lb of reinforcing bars.

   Table 4-8 shows prices for bending extras per 100 lb of reinforcing bars.

   Table 4-9 shows cost of placing reinforcing bars per ton.

3. Concrete—Use local ready-mix price per cu yd

4. Placing Concrete (labor and equipment per cu yd)
   Example 4-2 shows cost of placing concrete on upper floors.
   Example 4-3 shows cost of placing floor slabs on grade.
   Example 4-4 shows cost of placing concrete in large footings.
   Example 4-5 shows cost of placing concrete in bridge piers.

5. Finishing Concrete
   Table 4-12 gives cost of finishing concrete surfaces.

6. Curing Concrete—Cost varies widely; see discussion in text.

Referring to the information given in the preceding tables and examples we can select values to fit our problem and construct Tables 4-1P through 4-9P to cover the several cost factors.

TABLE 4-1P
Sizes of Members and Reinforcing

Item	Location	Size	Length or depth	Reinforcing
400	Footing	$8' \times 8' \times 2'$		10 #8 bars each way in bottom
410	Columns	$18'' \times 18''$	33'-0''	16 #9 column bars 16 #9 dowels $\times$ 5' −0'' 10 sets of #3 hoops each story
420	1st floor	$41' \times 20'$	0'-5''	#3 bars @ 12'' cts. each way
421	2nd, 3rd, and 4th floors	$41' \times 20'$	0'-6''	#4 bars @ 10'' cts. short way in top of slab #4 bars @ 14'' cts. short way in bottom #3 bars @ 12'' cts. long way in top and bottom
430	Long beams	$3'-0'' \times 1'-6''$	38'-0''	6 #10 bars and #5 stirrups $\times$ 7'-6'' @ 1'-6'' cts.
431	Short beams	$2'-6''* \times 1'-0''$	17'-0''	4 #7 bars and #4 stirrups $\times$ 6'-0'' @ 1'-3'' cts.

*Top of slab to bottom of beam.

TABLE 4-2P
Dimensions of the Part of a Member That Will Contain One Cu Yd of Concrete

Item	Location	Computation	Section	Contact area in 100 sq ft units
420	1st-floor slab	$X^2 \times 0.417 = 27$ $X = 8.05'$	$8.05' \times 8.05' \times 0.417'$	0.22*
421	2nd-, 3rd-, and 4th-floor slabs	$X^2 \times 0.5 = 27$ $X = 7.35'$	$7.35' \times 7.35' \times 0.5'$	$(7.35)^2/100 = 0.54$
430	Long beams	$X(2.5 \times 1.5) = 27$ $X = 7.21'$	$7.21' (2.5' \times 1.5')$	$7.2 \times 6.5 /100 = 0.47$
431	Short beams	$X(2 \times 1) = 27$ $X = 13.5'$	$13.5' (2' \times 1')$	$13.5 \times 5/100 = 0.68$
410	Columns	$X(1.5 \times 1.5) = 27$ $X = 12'$	$12'(1.5' \times 1.5')$	$12 \times 6/100 = 0.72$

*For the first-floor slab, which is on grade, forms are needed only for the turned-down portion around periphery of slab, which will be: $[2(2' + 1.58') (41 + 20') /100 \text{ sq ft}] \div 20$ cu yd = 0.22.

TABLE 4-3P
Lumber (fbm) and Hardware (lb) per Cu Yd of Concrete

		1	2	3	1 × 2	1 × 3
Item	Location	Contact surface in units of 100 sq ft per cu yd	Fbm per 100 sq ft of contact surface	Hdw per 100 sq ft of contact surface	Fbm of forms per cu yd	Hdw per cu yd
400	Footing					
410	Columns	0.72	200	8	144	6
420	1st floor	0.22	195	7	43	2
421	2nd, 3rd, and 4th floors	0.54	195	7	105	4
430	Long beams	0.47	400	10	188	5
431	Short beams	0.68	400	10	272	7
For quantities shown above see		4-2P	4-1	4-1		

Note: Use lumber price of $175 per 1,000 fbm and hardware price of $0.30 per lb.

TABLE 4-4P
Labor Hours for Forms per Cu Yd of Concrete

		1	2	3	4	1 × 2	1 × 3	1 × 4
			Labor hours per 100 sq ft of contact surface			Labor hours per cu yd concrete		
Item	Location	Contact surface in units of 100 sq ft per cu yd	To assemble	To erect	To strip and clean	To assemble	To erect	To strip and clean
400	Footing							
410	Columns	0.72	5	3	3	3.6	2.2	2.2
420	1st floor	0.22	5	3	3	1.1	0.7	0.7
421	2nd, 3rd, and 4th floors	0.54	5	3	3	2.7	1.6	1.6
430	Long beams	0.47	8	3	3	3.8	1.4	1.4
431	Short beams	0.68	8	3	3	5.4	2.1	2.1
For quantities shown above see		4-2P	4-2	4-2	4-2			

Note: Use $6.95 per hr average form labor (see first part of chapter). For equipment in carpenter shop use $2.00 per 100 sq ft of contact area.

TABLE 4-5P
Pounds of Reinforcement per Cu Yd of Concrete

Bar sizes			#3	#4	#5	#7	#8	#9	#10
Weight of bars in lb per lin ft			0.376	0.667	1.043	2.044	2.67	3.4	4.303
Item	*Location*	*Weight computation*							
400	Footings	20 bars × 7.67' × 2.67#/4.74 cu yd*					87		
410	Column bars	16 bars × 12'† × 3.4# × 1.19						777	
		(to allow for splice)							
	dowels	16 bars × 5.0' × 3.4# × 12'/33'						99	
	hoops	(4.5' + 5.83')0.376# × 12'/1.5'	31						
420	1st floor	2(8.05'† × 8.05'†/1')0.376#	49						
421	2nd, 3rd, 4th	(7.35'† × 7.35'†/0.83') 0.668#		43					
	floors	(7.35' × 7.35'/1.17') 0.668#		31					
		2(7.35' × 7.35'/1) 0.376#	41						
430	Long beams	6 bars × 4.303# × 7.2'†							186
	Stirrups	(7.2'/1.5') 1.043# × 7.5'			38				
431	Short beams	4 bars × 2.44# × 13.5'†					110		
	Stirrups	(13.5'/1.25') 0.667# × 6.0'		43					

*Cubic yards in one footing.

†See Table 4-2P for lengths of section to make 1 cu yd of concrete.

TABLE 4-6P
Cost of Reinforcement per 100 Lb

See Table 4-5P for lb of reinforcement per cu yd of concrete. See quantity sheet for cu yd of concrete for each item	Reinforcing bar sizes			
	#3	#4	#5	#6-11
lb cu yd lb cu yd lb cu yd #3 bars–(41 × 45.6) + (49 × 20) + (31 × 11) #4 bars–(43 × 45.6) + (31 × 45.6) + (43 x 18.9) #5 bars–(38 x 31.7) #6-11–(110 × 18.9) + (186 x 13.7) + (777 × 11) +(87 × 19) + (99 × 11) Total of all bars = 27,848/2,000 = 13.9 tons	3,191 lb	4,188 lb	12,505 lb	19,264 lb
Base price Richmond, Va. (Table 4-4)	$ 9.20	$ 9.20	$ 9.20	$ 9.20
Freight to Raleigh, N. C.	0.50	0.50	0.50	0.50
Quantity extra (Table 4-5)	1.25	1.25	1.25	1.25
Size extra (Table 4-6)	1.40	1.00	0.70	0.50
Detailing (Table 4-7)	1.15	1.15	1.15	1.15
Bending (Table 4-8)	2.25	1.50	1.50	1.50
Delivery to job	0.20	0.20	0.20	0.20
Chairs, spacers, ties etc.	0.50	0.50	0.50	0.50
Total	$16.45	$15.30	$15.00	$14.80
Suppliers' overhead and profit–12%	1.97	1.84	1.80	1.78
Materials: *Total cost per 100 lb of reinforcing bars delivered to job site (treat as material cost)	$18.42	$17.14	$16.80	$16.58
Labor Placing reinforcing bars by rodbuster, cost per 100 lb: #6 bars and larger @ $101.28 per ton/20† #5 bars and smaller @ $151.92 per ton/20†	$ 7.60	$ 7.60	$ 7.60	$ 5.06

Equipment:
Add $1.00 per cu yd of concrete for lifting reinforcing
bars with crane to each floor.

*This price is treated as a material cost (even though it
includes some labor cost) because the subcontractor has
already paid the taxes and insurance on this labor.
†See Table 4-9

TABLE 4-7P
Cost of Placing Concrete per Cubic Yard

Item	Reference	Labor	Equipment
420	See Example 4-3	$6.13	$0.85
410, 421, 430, 431	See Example 4-2	9.82	0.85
400	See Example 4-4	3.95	0.85

TABLE 4-8P
Finishing (Troweling Floors) per Cu Yd

(See Table 4-12)

Item					Labor
420 (8.05′ × 8.05′)	2 hr/100 sq ft	@	$6.93 per hour	=	$8.98
421 (7.35′ × 7.35′)	2 hr/100 sq ft	@	$6.93 per hour	=	$7.48

TABLE 4-9P
Curing Concrete per Cu Yd

410	Keeping forms wet (see discussion in text)	$0.50
420	Using membrane	0.25
421	Using membrane	0.20
430	Keeping forms wet	0.50
431	Keeping forms wet	0.50

Using information on Tables 4-1P through 4-9P we can fill out the Summary sheets for the quantity and unit price for each concrete item.

Because we are using the unit price method of estimating we must determine the pounds of reinforcing steel and fbm of formwork required per cu yd of concrete. And in order to estimate the weight of reinforcing bars per cu yd of a slab it is necessary to know the dimensions of the slab which will make one cu yd. From this piece of slab the contact area required for computing the formwork per cu yd can also be determined. This same information is also required for the columns, beams, and other sections. Tables 4-1P to 4-9P give this information.

The Quantity Takeoff sheets, the Summary and Unit Cost sheets, the Direct Cost sheets, and the Overhead and Profit sheets for Example 4-6 will now be shown.

QUANTITY TAKEOFF—Example 4-6

Estimate by _____ Date _____ Ckd. by _____ Date _____

Item	Identity	Location	Quantity Computations	Total	Units
400	Concrete	Footings	$4(8 \times 8 \times 2)/27$	19	cu yd
410	Concrete	Columns	$4(1.5 \times 1.5 \times 33)/27$	11	cu yd
420	Concrete	1st-floor slab	$(41 \times 20)0.42/27 + 2(41 + 20) \times 1 \times 1.58 /27$	20	cu yd
421	Concrete 2-3-4	Floor slabs	$3(41 \times 20)0.5/27$	45.6	cu yd
430	Concrete	Long beams	$3 \times 2(2.5 \times 1.5 \times 38) /27$	31.7	cu yd
431	Concrete	Short beams	$3 \times 5(2 \times 1. \times 17) /27$	18.9	cu yd

SUMMARIES & UNIT COSTS—Example 4-6

Estimate by _____ Date _____ Ckd. by _____ Date _____

Item No.	Identity & Cost Source	Computation of Unit Costs	Total	Units
400	Footings		19	cu yd
		Cost per cu yd delivered = $ 20.00*		
	Tables 4-5P, 4-6P	Reinf. (87#/100) $16.58 = 14.43		
		Material cost $ 34.43		
	Tables 4-5P, 4-6P	Placing reinf. (87#/100) $5.06 = $ 4.40		
	Table 4-7P	Placing conc. = 3.95		
		Labor cost $ 8.35		
	Table 4-7P	Lifting conc.		
		Equipment cost = $ 0.85		
410	Columns	(Figure 3 uses of forms)	11	cu yd
		Cost per cu yd delivered = $ 20.00		
	Table 4-3P	Forms (144 fbm/3) x 0.175 = 8.40		
	Table 4-3P	Hdw. -- 6# x $0.30 = 1.80		
	Tables 4-5P, 4-6P	Reinf. [(777# + 99#) /100]$16.58 = 145.24		
		Reinf. (31#/100 $18.42 = 5.71		
		Material cost = 181.15		
	Table 4-4P	Assemble forms (3.6 hr/3) $6.95 = $ 8.34		
	Table 4-4P	Erect, strip, clean - 4.4 hr x $6.95 = 27.80		
	Table 4-5P, 4-6P	Placing reinf. [(777# + 99#)/100] x $5.06 = 44.33		
	Table 4-5P, 4-6P	Placing reinf. (31#/100) $7.60 = 2.36		
	Table 4-7P	Placing conc. = 9.82		
	Table 4-9P	Curing conc. = .50		
		Labor cost $ 93.15		
	Table 4-6P	Lifting reinf. bars = $ 1.00		
	Table 4-7P	Lifting conc. = 0.85		
	Table 4-4P	Carpenter shop - $0.72 x $2.00 = 1.44		
		Equipment cost $ 3.29		
420	1st Floor	(Figure 3 uses of forms)	20	cu yd
		Cost per cu yd delivered = $ 20.00		
	Table 4-3P	Forms (43fbm/3) x $0.175 = 2.51		
	Table 4-3P	Hdw. -- 2# x $0.30 = 0.60		
	Table 4-5P, 4-6P	Reinf. (49#/100) $18.42 = 9.03		
		Material cost $ 32.14		
	Table 4-4P	Assemble forms (1.1 hrs/3) $6.95 = $ 2.55		
	Table 4-4P	Erect, strip, clean - 1.4 hr x $6.95 = 9.73		

SUMMARIES & UNIT COSTS — Example 4-6 (continued)

Estimate by _____ Date _____ Ckd. by _____ Date _____

Item No.	Identity & Cost Source	Computation of Unit Costs		Total	Units
	Tables 4-5P, 4-6P	Placing reinf. (49#/100)$7.60	= $ 3.72		
	Table 4-7P	Placing conc.	= 6.13		
	Table 4-8P	Finishing conc.	= 8.98		
	Table 4-9P	Curing conc.	= 0.25		
		Labor cost	$ 31.36		
	Table 4-6P	Lifting steel	= $ 1.00		
	Table 4-7P	Lifting conc.	= 0.85		
	Tables 4-2P, 4-4P	Carpenter shop 0.22 x $2.00	= 0.44		
		Equipment cost	$ 2.29		
421	2nd, 3rd, and 4th floors	(Figure 3 uses of forms)		45.6	cu yd
		Cost per cu yd delivered	= $ 20.00		
	Table 4-3P	Forms (105 fbm/3) x $0.175	= 6.12		
	Table 4-3P	Hdw. - 4# x $ .30	= 1.20		
	Tables 4-5P, 4-6P	Reinf. (74#/100)$17.14	= 12.68		
	Tables 4-5P, 4-6P	Reinf. (41#/100)$18.42	= 7.55		
		Material cost	$ 47.55		
	Table 4-4P	Assemble forms (2.7 hr/3)$6.95	= $ 6.26		
	Table 4-4P	Erect, strip, clean - 3.2 hr x $6.95	= 22.24		
	Tables 4-5P, 4-6P	Placing reinf. (115#/100)$7.60	= 8.74		
	Table 4-7P	Placing conc.	= 9.82		
	Table 4-8P	Finishing conc.	= 7.48		
	Table 4-9P	Curing conc.	= 0.20		
		Labor cost	$ 54.74		
	Table 4-6P	Lifting steel	= $ 1.00		
	Table 4-7P	Lifting conc.	= 0.85		
	Tables 4-2P, 4-4P	Carpenter shop — 0.54 x $2.00	= 1.08		
		Equipment cost	$ 2.93		
430	Long beams	(Figure 3 uses of forms)		31.7	cu yd
		Cost per cu yd delivered	= $ 20.00		
	Table 4-3P	Forms (188 fbm/3) $.175	= 10.96		
	Table 4-3P	Hdw. - 5# x $0.30	= 1.50		
	Tables 4-5P, 4-6P	Reinf. (186#/100) $16.58	= 30.84		
	Tables 4-5P, 4-6P	Reinf. (38#/100) $16.80	= 6.38		
		Material cost	$ 69.68		
	Table 4-4P	Assemble forms (3.8 hr/3) $6.95	= $ 8.89		
	Table 4-4P	Erect, strip, clean — 2.8 hr x $6.95	= 19.46		
	Tables 4-5P, 4-6P	Placing reinf. (186#/100) $5.06	= 9.41		
	Tables 4-5P, 4-6P	Placing reinf. (38#/100) $7.60	= 2.89		
	Table 4-7P	Placing conc.	= 9.85		
	Table 4-9P	Curing conc.	= .50		
		Labor cost	$ 51.00		

Estimate by _____ Date _____ Ckd. by _____ Date _____

Item No.	Identity & Cost Source	Computation of Unit Costs			Total	Units
	Table 4-6P	Lifting reinf. bars	=	$ 1.00		
	Table 4-7P	Lifting conc.	=	0.85		
	Tables 4-2P, 4-4P	Carpenter shop — 0.47 x $2.00	=	0.94		
		Equipment cost		$ 2.79		
431	Short beams	(Figure 3 uses of forms)			18.9	cu yd
		Cost per cu yd delivered	=	$ 20.00		
	Table 4-3P	Forms (272 fbm/3) $0.175	=	15.86		
	Table 4-3P	Hdw. — 7# x $0.30	=	2.10		
	Tables 4-5P, 4-6P	Reinf. (110#/100) $16.58	=	18.24		
	Tables 4-5P, 4-6P	Reinf. (43#/100) $17.14	=	7.37		
		Material cost		$ 63.57		
	Table 4-4P	Assemble forms (5.4 hr/3) $6.95	=	$ 12.51		
	Table 4-4P	Erect, strip, clean — 4.2hr x $6.95	=	29.19		
	Tables 4-5P, 4-6P	Placing reinf. (110#/100) $5.06	=	5.57		
	Tables 4-5P, 4-6P	Placing reinf. (43#/100) $7.60	=	3.27		
	Table 4-7P	Placing conc. — $9.85 x 3/4	=	7.45		
	Table 4-9P	Curing conc.	=	0.50		
		Labor cost		$ 58.49		
	Table 4-6P	Lifting reinf. bars	=	$ 1.00		
	Table 4-7P	Lifting conc.	=	0.85		
	Tables 4-2P, 4-4P	Carpenter shop .68 x $2.00	=	1.36		
		Equipment cost		$ 3.21		

*$20.00 per cu yd has been assumed as a national average cost of ready-mix concrete and includes sales tax and delivery charges. The sales tax is almost universal now and much of the concrete has to be delivered at distances necessitating a charge for the service. The cost of concrete delivered to the job site must be determined for each separate job.

DIRECT COSTS—Example 4-6

Estimate by _____ Date _____ Ckd. by _____ Date _____

Item No.	Identity & Location	Quantity		Unit Cost Each			Total Cost Each			Total Cost
		No.	Unit	Equip.	Mat'l.	Labor	Equip.	Mat'l.	Labor	
400	Conc. foot.	19	cu yd	$0.85	$34.43	$ 8.35	$ 16.00	$ 654	$ 159	$ 829
410	Conc. cols.	11	cu yd	3.29	181.15	93.15	36.00	1,993	1,025	3,054
420	1st-floor conc. slabs	20	cu yd	2.29	32.14	31.36	46.00	643	627	1,316
421	2nd, 3rd, and 4th floor conc. slabs	45.6	cu yd	2.93	47.55	54.74	134.00	2,168	2,496	4,798
430	Long conc. beams	31.7	cu yd	2.79	69.68	51.00	88.00	2,209	1,617	3,914
431	Short conc. beams	18.9	cu yd	3.21	63.57	58.49	61.00	1,201	1,105	2,367
	Direct costs						$381.00	$8,868	$7,029	$16,278

OVERHEAD & PROFIT—Example 4-6

Estimate by _____ Date _____ Ckd. by _____ Date _____

Item No.	Class of Expense	Computations of Overhead Expense	Total Cost
1600	Gen. overhead (% direct cost)	0.02 x $16,278	$ 326
	Job overhead	Assume 30 days for job	
1700	Int. on operating capital	10% x $16,278/2 x 0.09 (int.)/12 months	6
1701	Superintendent's salary	1/4 for job = $10,000 x 1/4 x 1/12 months	208
1702	Supt. pickup truck - rental	1/4 for job = 100 x 1/4	25
	do. operating cost	1/4 for job = 3,600 miles x 1/4 x 10¢ per mile	90
1703	Job trucks - rental		
	do. operating cost		
	do. wages - driver's		
1704	Lifting equipment	(Included in equipment)	
	do. operating cost	(Included in equipment)	
	do. wages - operators	(Included in labor)	
1705	Job office - rental		50
	do. salaries		
	do. supplies		33
1706	Utilities and connections	1/4 of job = $200 ÷ 4	50
1707	Social Security	5.85% of $7,029 x 90%*	370
1708	Workmen's Compensation	5% of $7,029	351
1709	Pub. Lia. & Prop. Damage	0.2% of $7,029	14
1710	Fed. & State Unemp. Ins.	3.2% x 45% x $7,029*	101
1711	Patents and royalties		
1712	Barricades		50
1713	Temporary toilets		50
1714	Cut and patch for trades		53
1715	Permits		50
1716	Protection adjacent prop.		
1717	Final cleanup		100
	Subtotal of computed overhead		$ 1,771
1718	Contingencies (% of computed overhead) 8% of $1,771		$ 142
	Total overhead		$ 1,913
	Direct cost from Direct Cost sheet		$ 16,278
	Subtotal (Total Overhead plus Direct Cost)		$ 18,191
	Profit 15% of $18,191		$ 2,729
	Total Cost		$ 20,920
	Performance bond $20,920 x 1/1000 x $10		$ 209
	Total amount of bid		$ 21,129
	* Assume average wage of $7.00 per hour		

In Homework Example 4-1 the weight of reinforcement per cubic yard of concrete will have to be computed using a different procedure than was employed in Example 4-6 because the bars are bent and are not continuous, and the reinforcing in columns varies from floor to floor. See Figure 4-5.

For slabs compute the cubic yards of concrete for one panel and divide into the pounds of reinforcing in the same panel to get pounds of steel per cubic yard.

For beams compute the cubic yards of concrete face to face of columns and divide into pounds of reinforcing in beams center to center of columns.

For columns compute the cubic yards of concrete in each section of column under consideration and divide into the weight of the column bars figured to include the bar splice at upper end.

For footings compute the cubic yards of concrete in footing and divide into the pounds of footing bars plus the column dowels.

The fbm of formwork per cubic yard of concrete can be determined as in Example 4-6—Tables 4-2P, 4-3P, and 4-4P.

For turned-down edge of slab on grade see Section B-B, Example 4-6.

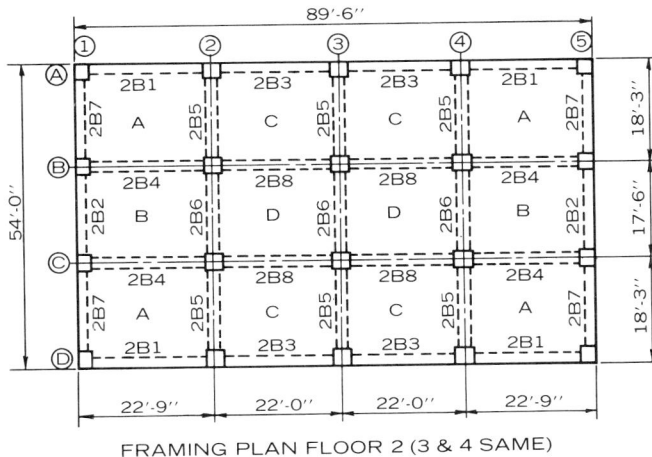

FRAMING PLAN FLOOR 2 (3 & 4 SAME)

FIGURE 4-5
Illustration for Homework Example 4-1
A.   Framing Plan Floor

TYPICAL BEAM & SLAB BENDING

2"
12" min.
6" slab
6"
2" clear for beams
¾" clear for slabs
L 1/4
L 2/4
L 1/7
L 1/5
L 2/5
L 1
L 2
D
O

TYP. EL. SHOWING STIRRUPS

2 #4 top for spandrel beams
stirrups

TYPICAL PLAN—SLAB "A"

2B6
2B8
2B2
Band "a"
Band "b"
Band "c"
2B4
Band "a"
Band "b"
Band "c"
2B5
Bend up 2 bars out of 3
2B7
2B3
2B1
LA/4
LA/4
L A
short span
LB/4
LB/4
L B
Long span

SECT. AT INTERIOR BEAM

3"
1½" clear
6"
D
W

SECT. AT SPANDREL BEAM

2 # 4 top continuous—Lap 1'-0" at splices
1½" clear
3"
6"
D
W

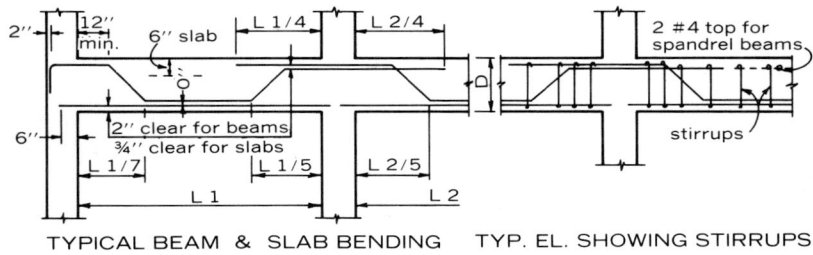

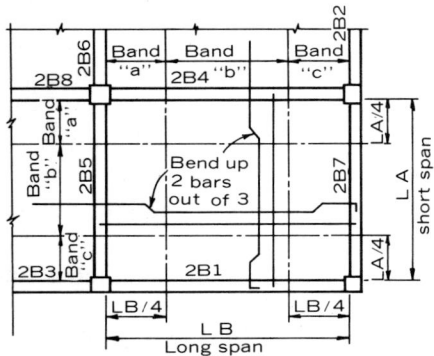

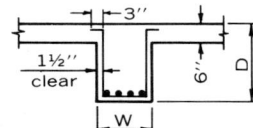

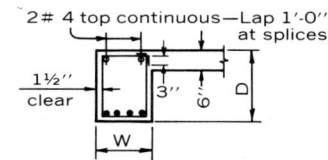

FIGURE 4-5 (continued)

B.   Typical Slab and Beam Sections

80   *Chapter Four*

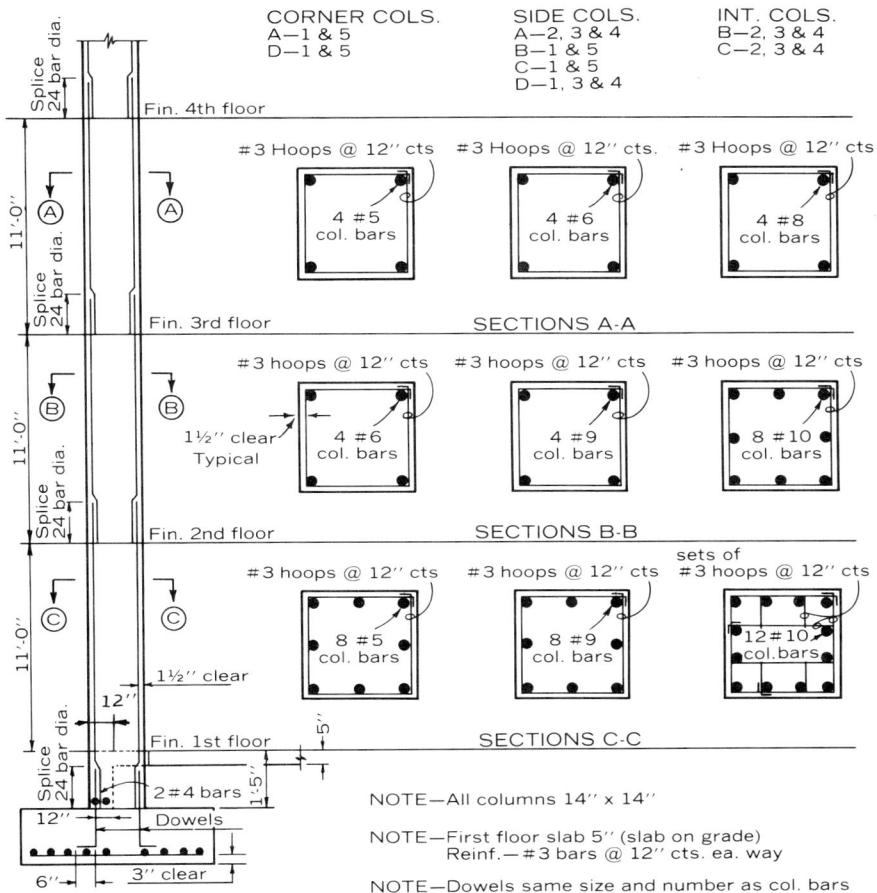

CORNER COLS.
A—1 & 5
D—1 & 5

SIDE COLS.
A—2, 3 & 4
B—1 & 5
C—1 & 5
D—1, 3 & 4

INT. COLS.
B—2, 3 & 4
C—2, 3 & 4

Splice 24 bar dia.

Fin. 4th floor

#3 Hoops @ 12'' cts   #3 Hoops @ 12'' cts   #3 Hoops @ 12'' cts

4 #5 col. bars   4 #6 col. bars   4 #8 col. bars

11'-0'' Splice 24 bar dia. (A)

Fin. 3rd floor

SECTIONS A-A

#3 hoops @ 12'' cts   #3 hoops @ 12'' cts   #3 hoops @ 12'' cts

1½'' clear Typical

4 #6 col. bars   4 #9 col. bars   8 #10 col. bars

11'-0'' Splice 24 bar dia. (B)

Fin. 2nd floor

SECTIONS B-B

sets of
#3 hoops @ 12'' cts   #3 hoops @ 12'' cts   #3 hoops @ 12'' cts

8 #5 col. bars   8 #9 col. bars   12 #10 col. bars

11'-0'' Splice 24 bar dia. (C)

1½'' clear

12''

5''

Fin. 1st floor

SECTIONS C-C

Splice 24 bar dia.

2 #4 bars

12'' Dowels

6'' 3'' clear

1'-5''

NOTE—All columns 14'' x 14''

NOTE—First floor slab 5'' (slab on grade)
Reinf.—#3 bars @ 12'' cts. ea. way

NOTE—Dowels same size and number as col. bars

C. Column Sections

FOOTINGS			
	At corner cols.	At side cols.	At int. cols.
Size	5'-6'' x 5'-6'' x 15''	7'-0'' x 7'-0'' x 20''	9'-0'' x 9'-0'' x 25''
Reinf.	7 #7 bars ea. way	8 #8 bars ea. way	10 #9 bars ea. way

NOTE—For computing lengths of hoops in columns and bent bars
in slabs see A.C.I. STAND. 315-65'' MANUAL OF STAND. PRACTICE
FOR DETAILING REINFORCED CONCRETE STRUCTURES.''

FIGURE 4-5 (continued)

D. Footing Schedule

BEAM SCHEDULE

Beam Mark	Size		Straight		Truss		Stirrups		
	W.	D.	No.	Size	No.	Size	No.	Size	Spacing ea. end
2B1	14	27	2	#7	2	#9	28	#3	4 @ 3, 4 @ 6, 3 @ 9, 3 @ 12
2B2	11	27	2	5	2	6	6	2	1 @ 9, 2 @ 12
2B3	14	27	2	6	2	9	28	3	4 @ 3, 4 @ 6, 3 @ 9, 2 @ 12
2B4	14	27	2	7	2	10	28	3	6 @ 3, 4 @ 6, 2 @ 9, 2 @ 12
2B5	14	20	2	7	2	9	20	3	2 @ 3, 4 @ 6, 2 @ 9, 2 @ 10
2B6	14	20	2	6	2	8	20	3	2 @ 3, 4 @ 6, 2 @ 9, 2 @ 10
2B7	11	27	2	5	2	7	6	2	1 @ 9, 2 @ 12
2B8	14	27	2	7	2	9	28	3	6 @ 3, 4 @ 6, 2 @ 9, 2 @ 12

E. Beam Schedule

SLAB SCHEDULE

Slab Mark	Direction	Bar Size	Bar spacing		
			Band "A"	Band "B"	Band "C"
A	N–S	#4	8″	6″	6″ W
	E–W	4	11	8½	8½ W
B	N–S	4	9	7	7 W
	E–W	4	11	8½	11
C	N–S	4	8	6	8
	E–W	4	13	10	10 W
D	N–S	4	9	7	9
	E–W	4	13	10	13

Note: Suffix "W" after bar spacing denotes band adjacent to wall.

FIGURE 4-5 (continued)

F. Slab Schedule

CHAPTER FIVE

Structural Steel

There are ten principal cost items which must be considered in computing the cost of structural steel erected in place and ready to be incorporated into the building:

1. Base cost of steel shapes and plates
2. Extras for the above items
3. Freight to fabricating plant
4. Detailing
5. Fabrication
6. Transportation to job site
7. Unloading and handling at job site
8. Erection
9. Riveting, bolting, and welding
10. Painting

All of these items will not apply to such prefabricated assemblies as bar joists, which are manufactured products, bought fob job site and which we will treat as material with only items 7, 8, 9, and 10 being considered.

BASE COST OF HOT ROLLED CARBON STEEL STRUCTURAL SHAPES

There is a base cost per 100 lb fob mill site to which must be added certain extras which will average 10 to 20% of the base price.

The base cost of steel fob mill site as of 1971 is shown in Table 5-1. Warehouse costs are much higher. The warehouse source, despite its higher prices, serves definite needs, such as orders from ornamental and other small shops who would find it difficult to buy in carload lots and also orders which are in large amounts that are required sooner than the mill can deliver. (However, it must be borne in mind that the warehouse will also add the usual extras to its base price.)

83

TABLE 5-1
Base Cost Hot Rolled Carbon Steel Structural
Shapes per 100 Lb fob

	Bethlehem, Pa.	Johnstown, Pa.	Lackawanna, N. Y.	Los Angeles, Calif.	Seattle, Wash.	So. San Francisco, Calif.	Date
Wide flange	$8.10		$8.10				8/5/71
Standard sections	8.10	$8.10	8.10	$8.20	$8.20	$8.20	8/5/71

EXTRAS

Quantity extras are charged for all orders below 4,000 lb and are figured for each individual size which is below this amount. By each individual size is meant each weight, gage, thickness, etc., of the structural piece. The quantity extras to be added to the base cost are shown in Table 5-2.

Mill extras are added to the cost of all plates and shapes to allow for the varied cost of rolling each item. Table 5-3 shows these costs.

Various sections used for columns and bridge members must be milled at the ends to achieve the required bearing. Table 5-4 shows the cost for this extra.

When required, beams may be cambered at the mill with the extra cost as shown in Table 5-5.

Rolled beam sections are sometimes split to form "Tee" sections. The extra costs for this operation are shown in Table 5-6.

There are also extras for special cutting, mechanical, chemical, and U.S. government requirements, and also for high-strength steel items. These items do not usually occur in the ordinary run of work; when they do, the prices should be thoroughly checked with the mill.

Steel plates have a different base price and different extras from structural shapes. Base prices and some extras are shown for hot rolled carbon plates.

TABLE 5-2
Quantity Extras

$ 5.00 per ton in the range of 1 to 2 tons
 15.00 per ton in the range of 1/2 to 1 ton
 45.00 per ton in the range of 0 to 1/2 ton

TABLE 5-3
Mill Extras per 100 Lb

Wide flange sections			
W4 × 13	$1.20	W14 × 30 to 38	$0.55
W5 × 16. and 18.5	1.00	W14 × 43 to 53	0.50
W6 × 8.5	1.70	W14 × 61 to 426	0.45
W6 × 12 and 16	1.60	W14 × 320 (col. core section)	0.45
W6 × 15.5 to 25	0.90	W16 × 26 and 31	0.75
W8 × 10	1.60	W16 × 36 to 50	0.55
W8 × 13 and 15	1.30	W16 × 58 to 96	0.45
W8 × 17 and 20	0.90	W18 × 35 and 40	0.65
W8 × 24 and 28	0.70	W18 × 45 to 60	0.50
W8 × 31 to 67	0.55	W18 × 64 to 114	0.45
W10 × 11.5	1.30	W21 × 44 and 49	0.60
W10 × 15 to 19	1.10	W21 × 55 to 142	0.45
W10 × 21 to 29	0.75	W24 × 55 and 61	0.60
W10 × 33 to 45	0.55	W24 × 68 to 160	0.45
W10 × 49 to 112	0.50	W27 × 84 to 177	0.45
W12 × 14	1.20	W30 × 99 to 132	0.45
W12 × 16.5 to 22	0.90	W30 × 172 to 210	0.50
W12 × 27 to 36	0.60	W33 × 118 to 152	0.50
W12 × 40 to 58	0.50	W33 × 200 to 240	0.55
W12 × 65 to 190	0.45	W36 × 135 to 194	0.50
W14 × 22 and 26	0.75	W36 × 230 to 300	0.55

Standard beam sections		Miscellaneous channels	
S3	$1.50	MC3 × 7.1 and 9	$1.40
S4	1.30	MC6 × 15.3 and 18	1.05
S5–S6–S7	1.10	MC × 6 × 12	1.10
S8	1.00	MC6 × 15.1 and 16.3	1.05
S10	0.85	MC7 × 19.1 and 22.7	1.00
S12 × 31.8 and 35	0.75	MC7 × 17.6	1.05
S12 × 40.8 and 50	0.80	MC8 × 21.4 and 22.8	0.90
S15 to S24	0.80	MC8 × 18.7 and 20	0.95
		MC9 × 23.9 and 25.4	0.90
Standard channels		MC10 × 24.9 and 28.3	0.85
C3	$1.40	MC10 × 21.9 and 25.3	0.85
C4	1.30	MC10 × 28.5 to 41.1	0.80
C5	1.20	MC12 × 30.9 to 37.0	0.70
C6 and C7	1.15	MC12 × 35.0 to 50	0.70
C8	1.10	MC13 × 31.8 to 50	0.70
C9	1.00	MC18 × 42.7 to 58	0.70
C10	0.90		
C12	0.80		
C15	0.70		

TABLE 5-3 (continued)

Standard angle sections

L8 × 8	$0.50	L9 × 4	$0.60	*L5 × 3	$0.70
*L6 × 6	0.65	L8 × 6	0.50	*L4 × 3 1/2	0.75
*L5 × 5	0.70	L8 × 4	0.65	*L4 × 3	0.80
*L4 × 4	0.70	L7 × 4	0.65	L3 1/2 × 3	0.80
L3 1/2 × 3 1/2	0.75	*L6 × 4	0.65	*L3 1/2 × 2 1/2	0.85
*L3 × 3	0.80	*L6 × 3 1/2	0.70	*L3 × 2 1/2	0.85
		*L5 × 3 1/2	0.70	*L3 × 2	0.90

*Except for special thicknesses as noted below

L6 × 6 × 5/16	$0.80		L5 × 3 1/2 × 1/4	$0.85
L5 × 5 × 5/16	0.90		L5 × 3 × 1/4	0.90
L4 × 4 × 1/4	0.85		L4 × 3 1/2 × 1/4	0.90
L3 × 3 × 3/16	0.90		L4 × 3 × 1/4	0.95
L6 × 4 × 5/16	0.80		L3 1/2 × 2 1/2 × 3/16	1.05
L6 × 4 × 1/4	0.80		L3 × 2 1/2 × 3/16	1.05
L6 × 3 1/2 × 1/4	0.85		L3 × 2 × 3/16	1.15

Bulb angles			*Rolled Tees*	
BL 10 × 3 1/2	$0.85		T3 × 6.7 and 7.8	$1.80
BL 9 × 3 1/2	0.85			
BL 8 × 3 1/2	1.00		*Standard Zees*	
BL 7 × 3 1/2	1.10		Z3	$1.10
BL 6 × 3 1/2	1.10		Z4	1.10
BL 5 × 4 1/2	1.10			
BL 5 × 3 1/2	1.10			
BL 4 × 3 1/2	1.30			

TABLE 5-4
Milling Extras per 100 Lb

Weight of section (Lb per ft)	One end			Two ends		
	Extra	Min	Max	Extra	Min	Max
10 to 50 incl.	$1.00	$10.00	$25.00	$1.25	$15.00	$30.00
Over 50 to 100 incl.	0.80	15.00	30.00	1.00	20.00	35.00
Over 100 to 200 incl.	0.60	20.00	35.00	0.80	25.00	40.00
Over 200 to 426 incl.	0.50	25.00	40.00	0.60	35.00	50.00
Over 426 inquire						

TABLE 5-5
Cambering Extras per 100 Lb

Weight of section (Lb per ft)	Extra	Per piece minimum	Per piece maximum
To 100 incl.	$0.75	$25.00	$40.00
Over 100 to 150 incl.	0.60	30.00	50.00
Over 150 to 200 incl.	0.40	35.00	60.00
Over 200 to 300 incl.	0.25	40.00	75.00

TABLE 5-6
Splitting Extras per 100 Lb*

Weight per foot of sect. before splitting (In lb)	Extras for Splitting Beams 5″ and Over to Produce Tees and Splitting Channels 5″ and Over to Produce Angles	
	Splitting by rotary shear (For each line of cut)	Splitting by flame cutting (For each line of cut)
Over 8 to 12 incl.	$1.60	$2.05
Over 12 to 15 incl.	1.40	1.85
Over 15 to 22 incl.	1.25	1.65
Over 22 to 45 incl.	0.80	1.30
Over 45 to 100 incl.	0.60	1.00
Over 100 to 150 incl.	0.50	0.80
Over 150 to 200 incl.	0.40	0.70
Over 200	— —	0.65

*The size extras for the beam or channel sections will also apply in addition to the splitting extras.
Note: Splitting by Rotary Shear: Beam and channel sections 18″ and under in depth with web thickness to 5/8″ inclusive carry price for rotary shear splitting.

Splitting by Flame Cutting: Beam and channel section with web thickness over 5/8″ must be split by flame cutting. Sections over 18″ in depth with web thickness to 5/8″ inclusive carry price for flame cutting.

BASE COST OF HOT ROLLED CARBON STEEL PLATES

There is a base cost per 100 lb fob mill site to which must be added certain extras. The base costs are shown in Table 5-7.

Tables 5-8 through 5-11 show the applicable extras to be added to the preceding base costs.

TABLE 5-7
Base Cost of Hot Rolled Carbon Steel Plates per 100 Lb

Johnstown, Pa.	Lackawanna, N.Y.	Sparrows Point, Md.	Burns Harbor, Ind.	Seattle, Wash.	Effective date
$7.40	$7.40	$7.40	$7.40	$7.50	3/1/71

TABLE 5-8
Item Quantity Extras per 100 Lb

(An item includes all the plates of a certain size and thickness: such as 12 plates 18 in. wide by 1 in. thick or 15 plates 20 in. wide by 2 in. thick, etc.)

Under 20,000 lb to 10,000 lb incl.	$0.10
Under 10,000 lb to 6,000 lb incl.	0.20
Under 6,000 lb to 4,000 lb incl.	0.30
Under 4,000 lb to 2,000 lb incl.	0.70
Under 2,000 lb to 1,000 lb incl.	1.25
Under 1,000 lb	2.00

TABLE 5-9
Size Extras per 100 Lb of Plates—Width and Thickness (Sheared and Gas Cut Edges)

Width in inches	Thickness in inches				
	Under 1/4	1/4 to 5/16 excl.	5/16 to 3/8 excl.	3/8 to 1/2 excl.	1/2 to 1 excl.
24 to 30 excl.	— —	$1.55	$1.35	$1.10	$1.05
30 to 36 excl.	— —	1.40	1.20	0.95	0.90
36 to 48 excl.	$1.60	1.35	1.10	0.85	0.80
48 to 60 excl.	1.45	1.20	0.95	0.70	0.65
60 to 80 excl.	1.45	1.05	0.85	0.60	0.45
80 to 90 incl.	1.50	1.05	0.80	0.55	0.35
Over 90 to 100 incl.	1.70	1.30	0.95	0.75	0.55
Over 100 to 110 incl.	1.90	1.50	1.20	0.95	0.75
Over 110 to 120 incl.	2.10	1.65	1.40	1.15	0.95
Over 120 to 130 incl.	2.30	1.85	1.60	1.35	1.15
Over 130 to 140 incl.	2.65	2.20	1.90	1.65	1.45
Over 140 to 152 1/2 incl.	3.00	2.45	2.20	1.95	1.70

There are also extras for special requirements such as: chemical; mechanical; circular and sketch plates; testing; heat treatment; surface finish; quality; certain specifications; etc. As in structural shapes these items do not usually occur in the ordinary run of work and can be obtained from the mill when required.

FREIGHT TO FABRICATING PLANT

The cost of freight from the mill to the fabricating plant must be added to the cost of the steel. This amount will vary considerably due to the type of material being moved as well as to the distance it is carried from the point of origin. For instance, long material requiring two cars will cost more to transport. This cost can vary from less than $0.50 to more than $2.00 per 100 lb. The estimator must determine these freight rates for his own locality.

TABLE 5-9 (continued)
Size Extras per 100 Lb of Plates—Width and Thickness (Sheared and Gas Cut Edges)

Width in inches	Thickness in inches				
	1 to 1 1/2 incl.	Over 1 1/2 to 3 incl.	Over 3 to 6 incl.	Over 6 to 12 incl.	Over 12
24 to 30 excl.	$1.10	$1.90	$1.95	$2.10	$2.20
30 to 36 excl.	1.00	1.70	1.75	1.90	2.00
36 to 48 excl.	0.90	1.50	1.60	1.80	1.90
48 to 60 excl.	0.75	1.30	1.45	1.60	1.70
60 to 80 excl.	0.65	1.10	1.30	1.40	1.50
80 to 90 incl.	0.55	0.95	1.15	1.25	1.35
Over 90 to 100 incl.	0.70	1.10	1.25	1.35	1.45
Over 100 to 110 incl.	0.80	1.15	1.30	1.45	1.55
Over 110 to 120 incl.	0.95	1.30	1.45	1.55	— —
Over 120 to 130 incl.	1.15	1.50	1.65	1.75	— —
Over 130 to 140 incl.	1.50	1.80	1.90	2.00	— —
Over 140 to 152 1/2 incl.	1.70	1.95	2.10	2.20	— —

Note: To all extras shown in Table 5-9, for plates over 1 1/2" thick, add $0.65 unless "Killed Steel" is implied or specified.

TABLE 5-10
Size Extras per 100 Lb of Plates—Width and
Thickness UM Edges

Width in inches	Thickness in inches				
	1/4 to 5/16 excl.	5/16 to 3/8 excl.	3/8 to 1/2 excl.	1/2 to 1 excl.	1 to 1 1/2 incl.
Over 8 to 12 excl.	$1.70	$1.50	$1.30	$1.25	$1.30
12 to 24 excl.	1.65	1.45	1.25	1.20	1.25
24 to 30 excl.	1.55	1.35	1.10	1.05	1.10
30 to 36 excl.	1.40	1.20	0.95	0.90	1.00
36 to 48 excl.	1.35	1.10	0.85	0.80	0.90
48 to 60 excl.	1.20	0.95	0.70	0.65	0.75
60	1.00	0.75	0.50	0.45	0.65

TABLE 5-11
Length Extras per 100 Lb of Plates

20'-0" to 50'-0"	No extra
Under 20'-0" to 8'-0" incl.	$0.05
Under 8'-0" to 5'-0" incl.	0.10
Under 5'-0" to 3'-0" incl.	0.20
Under 3'-0" to 2'-0" incl.	0.30
Under 2'-0" to 1'-0" incl.	0.50
Under 1'-0"	2.00
Over 50'-0" to and including 60'-0"	0.10
Over 60'-0" to and including 80'-0"	0.30
Over 80'-0" to and including 90'-0"	0.35
Over 90'-0" to and including 100'-0"	0.45
Over 100'-0" $0.50 + $0.10 for every additional 5'-0" or fraction thereof	

Note: Length extras apply on all thicknesses, whether plates
are sheared or gas cut.

TABLE 5-10 (continued)

Width in inches	Thickness in inches			
	Over 1 1/2 to 3 incl.	Over 3 to 6 incl.	Over 6 to 12 incl.	Over 12
Over 8 to 12 excl.	$1.50	$1.65	$1.90	$2.00
12 to 24 excl.	1.45	1.60	1.85	1.95
14 to 30 excl.	1.30	1.50	1.70	1.80
30 to 36 excl.	1.20	1.35	1.55	1.65
36 only	1.10	1.30	1.55	1.65
Over 36 to 48 excl.	1.10	1.30	1.55	1.65
48 to 58 incl.	1.00	1.20	1.40	1.50

Note: To all extras shown in Table 5-10, for plates over 1 1/2″ thick, add $0.65 unless "Killed Steel" is implied or specified.

DETAILING

Fabricating shops will usually quote the contractor on the structural steel for any job delivered to the job site and the estimator should always avail himself of this service on all jobs of any size. However, it is essential that he should know the mechanics of arriving at this figure.

The cost of detailing steel varies with the type of structure. Table 5-12 shows some approximate costs for this item.

FABRICATION

The cost of fabrication, like the cost of detailing, varies with the type of structure. Since the cost is based on weight it is evident that the heavier the sections the less the cost per pound will be because it takes very little more time to work the heavier material while the tonnage handled is greatly increased. So, heavy sections, plate girders, and base plates above 2″ in thickness will cost much less than light trusses. Also, shapes that are only punched will cost less to fabricate than the same pieces to which end connections must be fastened.

TABLE 5-12
Approximate Costs per Ton for Steel Detailing

Framing for office and storage buildings, hotels, etc.	$ 25.00	to	$ 55.00
Mill and factory bldgs. with cols., lintels	40.00	to	85.00
Trusses and craneways			
Framing for churches, museums, theaters, auditoriums, etc.	50.00	to	110.00
Plate girders	20.00	to	50.00
Highway trusses (per lin ft)	20.00	to	30.00
Roof trusses (per 24″ × 36″ sheet)	125.00	to	225.00

Table 5-13 shows some approximate fabrication prices.

Overhead and profit of the fabricator must be added to these costs because the product comes to the contractor as a finished material. An average overhead could be 10% of all costs and an average profit 10% of the total including overhead. Then this price will be treated as a material cost since the steel-fabricated figure already includes all labor costs and payroll taxes.

TABLE 5-13
Some Approximate Costs of Fabricating Structural Steel

	Cost per 100 lb
Beams and channels punched only	
to 10″	$1.50 - $2.40
12″ to 18″	1.30 - 1.90
over 18″	1.00 - 1.40
Beams and channels with end connections	
to 10″	2.80 - 3.80
12″ to 18″	2.30 - 3.30
over 18″	1.80 - 2.80
Beams framed with plates and angles	
to 10″	3.60 - 5.50
12″ to 18″	3.00 - 4.40
over 18″	2.30 - 3.20
Wide flange columns to 6″	5.00 - 7.00
6″ to 8″	4.00 - 6.00
over 8″	3.00 - 5.00
Built-up girders	4.00 - 6.50
Trusses	
under 1,500 lb	7.00 - 10.00
1,500 lb to 3,000 lb	6.00 - 9.00
over 3,000 lb	4.50 - 6.50
Base plates, anchor bolts, etc.	2.00 - 3.00

TRANSPORTATION TO JOB SITE

The cost of moving the fabricated steel from the shop to the job site will vary with the bulkiness of the material, which will limit the weight which a truck can accommodate, and also with the length of the haul. This cost may vary from under $0.35 to $0.60 or more per 100 lb.

To show the method of computing cost of delivering fabricated steel to the job assume:

1. Haul distance of 15 miles
2. 10-ton-capacity diesel truck @ $1,260 per month rental
3. Average speed of round trip is 20 mph
4. Driver @ $6.20 per hour

Rental cost of truck per hour = $1,760 ÷ 176	=	$10.00
Operating cost (See Example 3-1) = $0.017 × 175 hp	=	2.98
Tire wear (See Example 3-2)	=	0.34
Total cost of truck per hour	=	$13.32
Cost of driver per hour	=	6.20
Total cost of driver and truck per hour	=	$19.52

Trip time will be:

30 miles @ 20 mph	=	90 min.
Loading and unloading time 2 × 30	=	60 min.
Lost time, waiting maneuvering, etc.	=	30 min.
		180 min. = 3 hr

Cost per load = 3 hours × $19.52 = $58.56

Since this is a service furnished to the contractor the fabricator will add some overhead and profit. Ten percent overhead plus 10% profit would be 1.21 × $58.56 = $70.86.

Then cost per ton to deliver fabricated steel to the job site would be 70.86/10 = $7.09 or per 100 lb would be 7.09/20 = $0.35.

UNLOADING AND HANDLING AT JOB SITE

The fabricated pieces must be unloaded at the job and either stored or shook out. To shake out means to place the pieces as closely as possible to the geographical position on the ground which they will occupy in the building so that the crane boom can raise the member up to its final location without having to swing through a wide arc or do a lot of shifting forward or backward. This makes for much more efficient use of time by the raising gang and thus a cheaper erection cost. Just as in the case of transportation the cost of handling will vary quite a lot, say from less than $0.30 to more than $0.70 per 100 lb.

ERECTION

The cost of erection will vary with the size and height of the building. As the building goes higher the price will rise for three reasons: (1) some time will be required to raise the steel to the upper floors in addition to the regular setting time; (2) the wages of the steel workers increase as the working height above the ground increases; and (3) more equipment and personnel are required.

The cost of erection will also vary with the weight of the members handled. Heavy members can be set just about as fast as much lighter ones and will therefore show a greater tonnage placed at the same cost.

If a structure is straight framing with a lot of repetition, the price will be lower per ton. On the other hand, if the building is complicated with little repetition, the cost per ton will rise.

Erection times and prices are best determined from the experience of the company for which the estimate is being prepared. Table 5-14 shows some average labor hours per ton merely to illustrate one procedure in estimating erection costs.

RIVETING AND BOLTING

After the steel is erected and plumbed it must be riveted or bolted or welded to hold it in its final position and transfer the load from beam to girders to columns. High-strength bolting has replaced riveting for most structural work for several reasons: a two-man bolting crew can handle as much tonnage as a four-man riveting gang; the bolts are tightened to a very high stress and will not loosen even under severe vibration, such as encountered in a railroad bridge, as will rivets; and a bolted job can be completed in 2/3 to 3/4 of the time required for riveted jobs. Welding will be discussed later.

Riveted, bolted, and welded connections in the field are done from a platform, called a "float," which is approximately 3 ft wide by 6 ft long and is hung from the steel skeleton by a rope at each corner. This float has to be moved to a new location after each connection is made so that the time of the work will vary according to the number of rivets, bolts, or welds at each connection.

TABLE 5-14
Average Labor Hours for Steel Erection Including
Temporary Bolting and Plumbing

Type of work	Labor hours per ton
Roof trusses	
up to 1,500 lb	7 - 8
1,500 to 3,000	6 - 7
3,000 to 6,000	4 - 6
Beams and channels	3 - 8
Columns	4 - 8
Purlins and braces	5 - 6
Plate girders	3 - 6
Average structural frame	
up to 4 stories	4 - 8
Churches, theaters, etc.	5 - 7
Mill buildings with roof trusses and siding	4 - 12
Transmission towers	15 - 28
Framing for power plants	10 - 15
Base plates, anchor bolts, etc.	1 - 3

A riveting gang is generally composed of a riveter, a bucker, a catcher, and a heater. The heater keeps his hand forge filled with the proper length of rivets for each connection heated to the temperature that makes them plastic enough so that the riveting gun can shape a head on the straight end while the cupped bucking tool holds the preformed head on the other end of the rivet tightly against the steel plate. If the operation is properly done with exactly the right length of rivet used, then the rivet, upon cooling, will shrink enough to hold the connected parts tightly in place. If the rivet does not tighten properly, it must be removed and another one installed. This again adds to the cost of the riveting. When this trouble occurs in bolting it is easier and less costly to correct by simply tightening up a little on the bolt. The forge is placed in a central location so that it can furnish the rivets to several different connections before having to be moved and set up again. In order to keep the rivet from cooling enough to lose its plasticity the heater employs a long pair of tongs to throw it to the catcher, who traps it in a steel cup and uses a short pair of tongs to pick it out of the cup and "stick" it into the hole which it must fill—all within seconds after it leaves the forge.

The number of high-strength field bolts or rivets for most work will vary in the proximity of 15 to 30 per ton.

TABLE 5-15
Approximate Labor Hours for Installing Field
Rivets and Bolts per 100

	Rivets	High-strength bolts
Trusses	6 - 12	4 - 8
Steel office buildings	10 - 14	6 - 9
Steel mill buildings	10 - 12	6 - 8
Towers, etc.	14 - 18	9 - 12

TABLE 5-16
Cost of High-Strength A325 Bolts and Washers
per 100

Length of bolt in inches	5/8" φ	3/4" φ	7/8" φ	1" φ	1 1/8" φ	1 1/4" φ
1 1/2	$14.15	$20.70	$32.75	$48.85		
2	15.30	22.35	34.95	51.60		
2 1/2	16.45	24.10	37.40	54.70	$94.80	
3	17.55	25.85	39.90	58.00	98.60	$117.40
3 1/2	19.15	26.95	41.35	59.85	102.40	123.05
4	20.30	28.65	42.60	63.00	106.15	128.75
4 1/2	21.45	31.00	44.95	66.20	109.95	134.40
5	22.65	32.70	47.20	69.40	113.70	137.05
5 1/2	23.80	34.40	49.50	72.60	117.45	141.20
6	25.45	36.70	52.75	77.50	121.25	145.40
6 1/2	37.65	47.00	61.30	86.15	123.35	149.45
7	39.00	48.80	63.45	89.05	125.50	153.70
7 1/2	40.25	50.55	65.55	91.95	129.30	157.90
8	41.60	52.30	67.55	94.75	132.85	162.10
Hardened washers per 100	$2.45	$3.55	$4.90	$6.00	$7.00	$8.75

Notes: Cost per 100 bolts in quantities of 10,000 to 20,000 lb will be as on table. Cost per 100 bolts in quantities of 5,000 to 10,000 lb will be approximately 5% more. Cost per 100 bolts in quantities under 5,000 lb will be approximately 11% more. Cost per 100 bolts in quantities over 20,000 lb will be approximately 7% less.

Usually one hardened washer is used under the element (nut or bolt head) turned in tightening.

The number of bolts ordered has considerable bearing on their price. When there are fewer bolts requested than the number required to fill a shipping container, then 20% must be added to the prices shown. Table 5-17 shows the number of bolts of each size and length required to fill the container.

The approximate time to install and tighten the fasteners is shown in Table 5-15.

The material costs for bolts, based on their lengths, are shown in Table 5-16. The length of bolts is determined using this procedure:

1. First determine the grip (thickness of the pieces to be joined in inches).
2. Then allow for the added lengths, due to nuts, as follows: for 5/8"ϕ bolts, add 7/8" to grip; for 3/4"ϕ, add 1"; for 7/8"ϕ, add 1 1/8"; for 1"ϕ, add 1 1/4"; for 1 1/8"ϕ, add 1 1/2"; for 1 1/4"ϕ, add 1 5/8".
3. Add 3/16" for each flat washer used.
4. Specify bolt lengths to the next highest 1/4".

TABLE 5-17
Container Quantities for A325 High-Strength Bolts

Length of bolt in inches	5/8" ϕ	3/4" ϕ	7/8" ϕ	1" ϕ	1 1/8" ϕ	1 1/4" ϕ
1 1/2	650	350	250	225		
2	550	350	250	200		
2 1/2	450	300	200	180	125	
3	400	250	175	150	100	80
3 1/2	350	225	175	125	100	80
4	300	200	150	115	90	70
4 1/2	250	175	125	110	90	65
5	200	150	125	100	80	60
5 1/2	200	150	110	95	70	55
6	175	135	100	95	70	55
6 1/2	150	125	100	75	60	50
7	150	125	90	70	60	45
7 1/2	140	110	85	70	50	40
8	140	110	75	60	50	40

The tables for the material costs of rivets are based on weights. To determine the weights of rivets:

1. Determine the grip (thickness of pieces to be joined) in inches.
2. Enter Table 5-18 with grip to determine length of rivet required.

3. Then enter Table 5-19 with length of rivet required to find weight of rivets per 1,000 pieces.
4. Use Tables 5-20 and 5-21 to find the price extras to be added to the base cost per 100 lb to get the final material cost of the rivets.

The average 1970 base cost of rivet stock is $16.00 per 100 lb fob plant, but the 1971 prices are only available upon application to the plant. There are several extras to be applied to this price as shown in Tables 5-20 and 5-21.

TABLE 5-18
Lengths of Rivets per Inch of Grip

Grip	Diameter in inches					Grip	Diameter in inches				
	1/2	5/8	3/4	7/8	1		1/2	5/8	3/4	7/8	1
1/2	1 5/8	1 7/8	1 7/8	2	2 1/8	2	3 1/2	3 1/2	3 5/8	3 3/4	3 7/8
5/8	1 3/4	2	2	2 1/8	2 1/4	2 1/8	3 5/8	3 5/8	3 3/4	3 7/8	4
3/4	1 7/8	2 1/8	2 1/8	2 1/4	2 3/8	2 1/4	3 3/4	3 7/8	3 7/8	4	4 1/8
7/8	2	2 1/4	2 1/4	2 3/8	2 1/2	2 3/8	4	4	4	4 1/8	4 1/4
						2 1/2	4 1/8	4 1/8	4 1/8	4 1/4	4 3/8
1	2 1/4	2 3/8	2 3/8	2 1/2	2 5/8	2 5/8	4 1/4	4 1/4	4 1/4	4 3/8	4 1/2
1 1/8	2 3/8	2 1/2	2 1/2	2 5/8	2 3/4	2 3/4	4 3/8	4 3/8	4 3/8	4 1/4	4 5/8
1 1/4	2 1/2	2 5/8	2 5/8	2 3/4	2 7/8	2 7/8	4 5/8	4 5/8	4 5/8	4 5/8	4 3/4
1 3/8	2 5/8	2 3/4	2 3/4	2 7/8	3						
1 1/2	2 7/8	3	3	3 1/8	3 1/4	3		4 3/4	4 3/4	4 7/8	5
1 5/8	3	3 1/8	3 1/8	3 1/4	3 3/8						
1 3/4	3 1/8	3 1/4	3 1/4	3 1/2	3 5/8						
1 7/8	3 1/4	3 3/8	3 3/8	3 5/8	3 3/4						

WELDING

As stated before, high-strength bolting has in most cases supplanted riveting as a means of securing structural steel members in place in buildings, towers, bridges, and most other structures. Even in the fabricating shops, bolting is taking precedence over riveting.

TABLE 5-19
Approximate Weight in Lb for Button-Head
Rivets per 1,000 Pieces

Length in inches	Diameter in inches					Length in inches	Diameter in inches				
	1/2	5/8	3/4	7/8	1		1/2	5/8	3/4	7/8	1
1 1/2	126	216	317	473	693	3 1/2	237	375	565	835	1125
1 5/8	133	226	333	495	720	3 5/8	244	385	581	857	1152
1 3/4	140	236	348	517	747	3 3/4	250	395	596	879	1179
1 7/8	147	246	364	539	774	3 7/8	257	405	612	901	1206
2	154	256	379	561	801	4	264	414	627	923	1233
2 1/8	161	266	395	583	828	4 1/8	271	424	643	945	1260
2 1/4	168	276	410	605	855	4 1/4	278	434	658	967	1287
2 3/8	175	286	426	627	882	4 3/8	285	444	674	989	1314
2 1/2	181	297	441	649	909	4 1/2	292	454	689	1011	1341
2 5/8	188	306	457	671	936	4 5/8	299	464	705	1033	1368
2 3/4	195	315	472	693	963	4 3/4	306	474	720	1055	1395
2 7/8	202	325	488	715	990	4 7/8	313	484	736	1077	1422
3	209	335	503	737	1017	5	319	494	751	1099	1449
3 1/8	216	345	519	759	1044						
3 1/4	223	355	534	781	1071						
3 3/8	230	365	550	813	1098						

Note: Weights have been rounded to the nearest pound per 1,000 pieces.

In some cases, welding is being used in building work both in the fabrication in the shop and in the erection in the field. When the connections are welded there is a saving of weight because the connection angles and plates are practically all eliminated. Also, where continuous beams or rigid frames are used in a structure there is a considerable saving in the size of the main members due to continuity. Here welding is almost a necessity since riveted or bolted connections would require a prohibitive amount of connection material and in addition the large number of connection holes would reduce the section and dictate larger main members in order to maintain the required net section.

TABLE 5-20
Quantity Extras per 100 Lb for Rivets

Quantities per item	Item quantity	Standard sizes
Extras apply to weight of any item under 2,000 lb regardless of weight of entire order	Under 100 lb	$10.00
	100 - 199 lb	5.00
	200 - 399 lb	2.75
	400 - 599 lb	1.50
	600 - 999 lb	1.00
	1,000 - 1,999 lb	0.50
Quantities per order	Total order	Any size
Based on total order for shipment to one destination in one lot	under 1,000 lb	$3.00
	1,000 - 1,999 lb	2.00
	2,000 - 4,999 lb	1.25
	5,000 - 9,999 lb	0.50
	10,000 - 19,999 lb	0.25
	20,000 lb and more	No extra

TABLE 5-21
Size Extras per 100 Lb for Rivets

	Diameter in inches	Length in inches	Extra
Standard sizes, button head rivets in standard steel grades in 1/8″ increments of sizes shown at right	1/2	1 1/4 to 3 1/2	$2.75
	5/8	1 1/4 to 3 1/2	2.00
	3/4	1 1/2 to 4	1.75
	7/8	2 1/2 to 5	1.75
	1	3 to 5	3.25

Note: All of the above extras will change for special grades of steel and nonstandard sizes of rivets.

The detailing in the office and the fabrication in the shop will usually require less work because the only holes necessary are those required for erection purposes to hold the members in place until they can be welded and also because connection angles and plates are not required.

These are the advantages of welding. However, there are some disadvantages. Although from 10 to 20% of overall material can be saved and some saving realized on detailing and fabricating, the erection requires a type of personnel with skills that are harder to recruit than are the iron workers to do the customary bolted construction. It is much more difficult to set up the equipment for and to maintain high-quality welding in the field than in the fabricating shop. When a project is of such size that it can assimilate the mobilizational time required, then an overall saving can be realized since the weight saving will not be offset by the increased erection cost.

There are many factors that control the cost of welding: the accuracy of the shop work; the size and type of the welds; the accessibility of the welds; the skill of the welders; and the ability of the designer to arrange connections that can be economically welded in both the fabricating shop and in the erection in the field.

Tables 5-22 and 5-23 give some approximate costs of welding per linear foot of welds. The tables assume that a welding gang will be composed of 2

TABLE 5-22
Approximate Costs per Linear Foot of Fillet Welds

Weld size	Pounds of weld rod per linear foot	Cost of material per linear foot	Cost of labor per linear foot
1/8″	0.1	$0.03	$ 0.16
3/16	0.2	0.06	1.32
1/4	0.3	0.10	1.98
5/16	0.4	0.13	2.64
3/8	0.5	0.16	3.30
1/2	0.7	0.22	4.62
3/4	1.4	0.42	9.24
1	2.4	0.77	15.84

TABLE 5-23
Approximate Costs per Linear Foot of Butt Welds

Thickness of plate	Pounds of weld rod per linear foot	Cost of material per linear foot	Cost of labor per linear foot
1/4	0.4	$0.13	$ 2.64
5/16	0.5	0.16	3.30
3/8	0.7	0.22	4.62
1/2	0.8	0.25	5.28
3/4	1.6	0.48	10.56
1	2.6	0.84	17.16

mechanics who will share a helper between them. The hourly cost will be $8.35 for the welder and $6.40/2 or $3.20 for the helper, making $11.55 total. Each welder is assumed to deposit 1.75 lb of weld rod per hour. The cost of the weld rod is assumed to be $0.18 per pound. The pounds of weld rod required per linear foot of weld carry an allowance for spatter loss and the cost per linear foot of the rod carries an allowance for electricity.

These tables show an approximate material and labor cost. The equipment costs will vary with the type of welding machines employed in the work. The cost of equipment can vary from $1.50 to $3.00 per ton of steel.

PAINTING

After erection the steel must be painted, unless it is to be encased in concrete. The cost per ton will vary with the weight of the section since a light section will have about the same area as a heavier piece of the same depth while accounting for much less tonnage. In the same manner, a truss will cost much more to paint than an equal weight of plate girder. Table 5-24, which shows approximate costs, is based on a paint costing $6.00 per gal and covering 400 sq ft. It is assumed that a man can clean and paint 160 sq ft per hour at a wage of $7.45 per hour. These costs will vary with each different area and the times should be checked with the work experience of the firm handling the project.

TABLE 5-24
Approximate Painting Costs per Ton

Material	Square ft per ton	Paint	Labor
3″ X 8″ angles	300 - 350	$4.50 - $5.25	$14.00 - $16.30
Standard channels	325 - 375	4.90 - 5.60	15.15 - 17.45
Beams	200 - 250	3.00 - 3.75	9.30 - 11.65
Girders	125 - 200	1.90 - 3.00	5.80 - 9.30
Columns	200 - 250	3.00 - 3.75	9.30 - 11.65
Roof trusses	350 - 400	5.25 - 6.00	16.30 - 18.60
Bridge trusses	200 - 250	3.00 - 3.75	9.30 - 11.65

TAKING MATERIAL FROM PLANS

In taking material from plans use lengths of beams from center to center of columns ignoring the space occupied by the columns and do the same with beams framing into girders. This allows for the wastage which always occurs due to many reasons. Make no allowances for clipping, milling, punching, or boring. Use rectangular dimensions for all plates except where it is obvious that a number of irregular plates can be cut from one large rectangular plate.

Ordinarily there is no material taken off for details such as connections, small miscellaneous pieces, etc. End brackets on spandrel beams that serve as wind bracing, stiffeners, angles, etc., are not considered detail. Table 5-25 gives some approximate percentages which may be added to take care of details.

TABLE 5-25
Approximate Percentage of Details

Columns	10 - 20
Beams	5 - 25
Built-up girders	10 - 15
Roof trusses	15 - 25

EXAMPLE 5-1

Find the cost of the steel frame for the building in Figure 5-1 erected in place and painted. The building will consist of 4 stories and a roof as shown. High-strength field bolts will be employed. Assume it will require 30 days to erect, bolt, and paint the structural steel.

The following order of work will be observed:

1. Compute an average hourly wage for the erection crew and for the bolting gang; next select the equipment required and then determine the rental and operating cost of this equipment.
2. Then take material from the plans, assign an item number, and record it on the Quantity sheets. The item numbers will be selected to cover each differently priced section shown in the table for fabricating costs because this will be the largest of the costs. It may be necessary to average one or more of the additional costs because they may not vary for the same items for which the fabricating costs vary.

COLUMN SCHEDULE

Columns	A1; A7; D1; D7	A2 to A6; D2 to D6 B1; B7; C1; C7	B2 to B6; C2 to C6
Roof line			
+51'-0" 2½" Tops of cols.	W 12 × 40	W 12 × 40	W 12 × 58
Fin. 3rd floor +38'-0"			
Col. splice			
Fin. 2nd floor 1'-6" +26'-0"	W 12 × 65	W 12 × 72	W 12 × 99
Fin. 1st floor +14'-0"			
Fin. bsmt. Top of base Pl. & bottom of cols. 1'-0" +0'-0"			
Col. base plates	18" × 1½" × 20"	19" × 1⅝" × 21"	22" × 2¼" × 24"

€ plan symm. about €

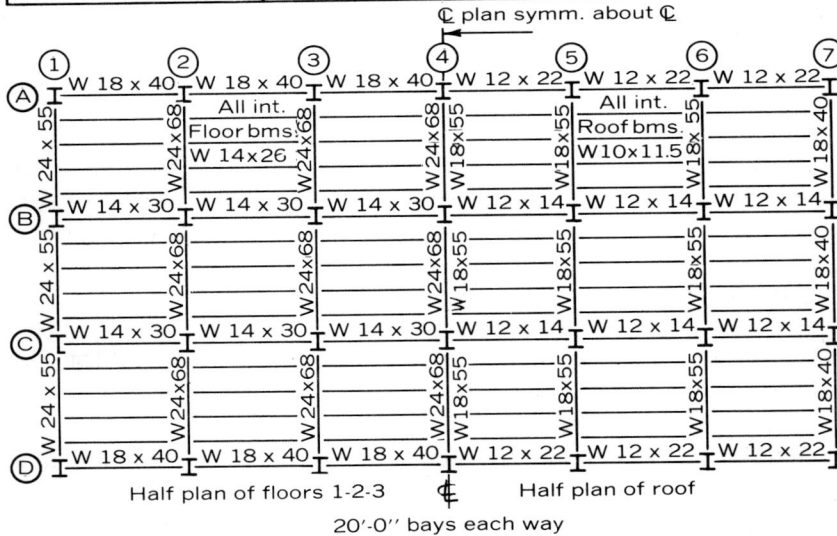

Half plan of floors 1-2-3   €   Half plan of roof

20'-0" bays each way

FIGURE 5-1
Column Schedule for Storage Building for Example 5-1

3. Next group together the material in the several item numbers on the Summary sheets and compute the cost for material, labor, and equipment for each separate item.

4. In the final operation carry the total quantities and unit costs for the several items to the Direct Cost sheets and make extensions. Then add the overhead and profit items, compute the performance bond, and add to the total.

The average hourly wage for a typical erection gang is computed as follows:

1 pusher @ $8.90 per hour	=	$ 8.90
2 connectors @ $8.55	=	17.10
1 man to hook up	=	8.55
1 tag line man	=	8.55
1 crane operator	=	8.20
1 oiler	=	6.05
		$57.35

Average hourly wage = $57.35/7 = $8.19. Round off to $8.20.

Bolting is usually done with two-man gangs equipped with an impact wrench and bucking tool. On an average job there will probably be three such gangs of two men each with a compressor and an engineer. The average hourly wage is computed as follows:

6 iron workers @ $8.55 per hour	=	$51.30
1 engineer @ $7.95 per hour	=	7.95
		$59.25

Average hourly wage for bolting gang = $59.25/7 = $8.46

The following equipment will be required:

1 30-ton diesel crane for one month*	=	$2,206
Operating cost—150 hp × $0.017 × 176 hr†	=	449
1 diesel-engine compressor for one month*	=	537
Operating cost for compressor	=	100
3 impact wrenches for one month*	=	279
Total equipment cost (using bolting)		$3,571

If rivets were used in place of high-strength bolts the labor and equipment costs would be computed as follows. Riveting is usually done with gangs

---

*From Appendix A.
† See Example 3-1.

composed of 4 steel workers each. Generally one compressor serviced by one enginer will be required for 3 such gangs. The average hourly wage is computed:

$$
\begin{array}{lll}
\text{12 iron workers @ \$8.55} & = & \$102.60 \\
\text{1 engineer @ \$7.95} & = & \underline{7.95} \\
& & \$110.55
\end{array}
$$

$$\text{Average hourly wage for riveting gang} = \frac{\$110.55}{13} = \$8.50$$

The following equipment will be required:

$$
\begin{array}{lll}
\text{1 30-ton diesel crane for one month*} & = & \$2,206 \\
\text{Operating cost—150 tons} \times \$0.017 \times 176 \text{ hr†} & = & 449 \\
\text{1 diesel-engine compressor for one month} & = & 537 \\
\text{Operating cost for compressor} & = & 100 \\
\text{3 riveting guns for one month} & = & 150 \\
\text{3 forges, tongs, etc.} & = & 75 \\
\text{Coke for forges} & = & \underline{15} \\
& & \$3,532
\end{array}
$$

It will be seen from these figures that the equipment cost for bolting is practically the same as for riveting.

The average hourly wage for bolting is the same as for riveting. However, an examination of Table 5-15 will show that the man hours required for installing high-strength bolts is considerably less than for rivets, thus providing a savings in labor cost when using bolting as well as an overall saving in job completion time.

The material cost for rivets is lower than for high-strength bolts. However, in the overall picture bolting is considered to be cheaper.

The material for Example 5-1 will now be taken from Figure 5-1 and shown on the Takeoff forms.

---

*See Appendix A.
†See Example 3-1.

QUANTITY TAKEOFF—Example 5-1

Estimate by _____   Date _____   Ckd. by _____   Date _____

Item	Identity	Location	Quantity Computations	Total	Units
				*	lb
543	Struc. steel	Roof beams	W10-(11.5# x 20') 72 x 1.10	18,205	
542	- - - - - -	- - - - - -	W12-(22# x 20') 12 x 1.05	5,544	
542	- - - - - -	- - - - - -	W12-(14# x 20') 12 x 1.05	3,528	
542	- - - - - -	- - - - - -	W18-(40# x 20') 6 x 1.05	5,040	
542	- - - - - -	- - - - - -	W18-(55# x 20') 15 x 1.05	17,325	
542	Struc. steel	Floor beams	W14-3(26# x 20') 72 x 1.05	117,915	lb
542	- - - - - -	- - - - - -	W18-3(40# x 20') 12 x 1.05	30,240	
542	- - - - - -	- - - - - -	W14-3(30# x 20') 12 x 1.05	22,680	
541	- - - - - -	- - - - - -	W24-3(55# x 20') 6 x 1.05	20,790	
541	- - - - - -	- - - - - -	W24-3(68# x 20') 15 x 1.05	64,250	
540	Struc. steel	Columns	W12-(99# x 28.5') 10 x 1.10	31,020	lb
540	- - - - - -	- - - - - -	W12-(58# x 23.3') 10 x 1.10	14,850	
540	- - - - - -	- - - - - -	W12-(72# x 28.5') 14 x 1.10	31,625	
540	- - - - - -	- - - - - -	W12-(40# x 23.3') 14 x 1.10	14,355	
540	- - - - - -	- - - - - -	W12-(65# x 28.5') 4 x 1.10	8,151	
540	- - - - - -	- - - - - -	W12-(40# x 23.3') 4 x 1.10	4,103	
544	Struc. steel	Base pls.	4 pls. 18" x 1 1/2" x 20"	612	lb
			14 pls. 19" x 1 5/8" x 21"	2,576	
			10 pls. 22" x 2 1/4" x 24"	3,370	
			Total =	416,179	lb
				or 208.1	tons

*Total weight plus the proper percentage for details

NOTE: Select an item number to cover each different fabricating
cost (since this is usually the largest single cost) and
average other costs when necessary. See Table 5-13.

Most of the beams in this structure will only have end connections so they will
be covered by that part of Table 5-13 captioned "Beams and channels with end
connections." The items were selected as shown below:

Item 543 - For 10" beams the range in table for fabrication cost per 100 lb
is $2.80 to $3.80 - We chose $3.30.

Item 542 - For 12" to 18" beams the range is $2.30 to $3.30 per 100 lb
We chose $2.80.

Item 541 - For 24" beams - range $1.80 to $2.80. We chose $2.30.

Item 540 - For 12" columns - range $3.00 to $5.00. We chose $4.00.

Item 544 - For base plates - range $2.00 to $3.00. We chose $2.00.

Estimate by _____ Date _____ Ckd. by _____ Date _____

Item No.	Identity & Cost Source	Computation of Unit Costs	Total	Units

540	Steel cols.
	Table 5-1
	" 5-2
	" 5-3 xx
	
	" 5-12
	" 5-13
	
	Table 5-16
	
	Table 5-24
	
	Table 5-14
	See text
	Table 5-15
	See text
	
	Table 5-24
	
	See equipment
	estimate
		W12 Cols. & beams + details
		Base cost fob mill per 100# = $ 8.10
		Quantity extra " " = 0.00
		Mill extra " " = 0.47
		Freight to fabricator = 0.50
		Steel detailing " " = 2.00
		Fabrication " " = 4.00
		Transportation to job per
		100# = <u>0.35</u>
		      Total $15.42
		Fab. overhead - 5% " " = <u>0.80</u>
		      Total $16.22
		Fab. profit - 5% " " = 0.83
		Fab. steel, job site per
		100# = $17.05
		Field bolts (20 bolts
		per ton - 7/8" x 3" avg
		$0.40 x 20/20* = 0.40
		Field paint $3. per ton /20* = <u>0.15</u>
		Materials per 100# = $17.60
		<u>Materials</u> per ton = * 20 x
		17.60 =$352.00
		
		Unloading & handling at
		job site - per ton $ 0.50
		Erection - 6 hr per ton @
		$8.25 per hr = $49.50
		Bolting - 20 bolts per ton
		@ 6 hr per 100 bolts @ 8.46
		per hr = 10.15
		Painting per ton = <u>9.30</u>
		<u>Labor</u> " " $69.45
		
		Equipment cost per ton
		$3571/208.1 tons = $17.15
			52	Tons

		xx The steel members having the same cost of fabrication were assembled in the same item. The other extras, where necessary, were averaged. As an ex- ample consider the mill extra for item 540: Table 5-3 shows the mill extra as follows:		

Estimate by _____ Date _____ Ckd. by _____ Date _____

Item No.	Identity & Cost Source	Computation of Unit Costs	Total	Units
		For W12 x 99 cols - $0.45 per 100# For W12 x 72 cols - $0.45 per 100# For W12 x 65 cols - $0.45 per 100# For W12 x 58 cols - $0.50 per 100# For W12 x 40 cols - $0.50 per 100# The amount of the mill extra for item 540 was proportioned according to the weights of the several members which averaged out to $0.47 per 100#.		

541	Steel beams
	 Table 5-1
	 " 5-2
	 " 5-3
	
	 " 5-12
	 " 5-13
		WF 24 Beams & details
		 Base cost fob mill per 100# = $ 8.10
		 Quantity extra " " = 0.00
		 Mill extra " " = 0.49
		 Freight to
		 fabricator " " = 0.50
		 Steel detailing " " = 1.25
		 Fabrication " " = 2.30
		 Transportation
		 to job " " = 0.35
		      Total per 100# = $12.49
		 Fab. overhead - 5% per 100# = 0.65
		      Total per 100# $13.14
		 Fab. profit - 5% " " = 0.68
		 Fab. steel, job site per 100#= $13.82
			42.5	Tons

	Table 5-16 Table 5-24	Field bolts (20 bolts per ton - 7/8 x 3" avg $0.40 x 20/20* = 0.40 Field paint $3. per ton/20* = 0.15 Materials per 100# $14.37 <u>Materials</u> per ton = 20 x 14.37 =$287.40		
	Table 5-14 See text Table 5-15 See text Table 5-24	Unloading & handling at job site per ton = $ 0.50 Erection - 5 hr per ton @ $8.25 per hour = $41.25 Bolting - 20 bolts per ton @ 6 hr per 100 bolts @ $8.46 per hr = $10.15 Painting per ton = 9.30 <u>Labor</u> " " $61.20		
	See equipment estimate	Equipment cost per ton $3571/208.1 tons = $17.15		

Estimate by _____ Date _____ Ckd. by _____ Date _____

Item No.	Identity & Cost Source	Computation of Unit Costs	Total	Units
542	Steel beams	W12&18 Beams & details	101.2	tons
	Table 5-1	Base cost fob mill per 100# = $ 8.10		
	Table 5-2	Quantity extra " " = 0.00		
	" 5-3	Mill extra " " = 0.70		
		Freight to fabricator per 100# = 0.50		
	" 5-12	Steel detailing " " = 1.50		
	" 5-13	Fabrication " " = 2.80		
		Transportation to job " " = 0.35		
		Total " " $13.95		
		Fab. overhead - 5% " " 0.75		
		Total " " $14.70		
		Fab. profit - 5% " " 0.80		
		Fab. steel, job site " " $15.50		
	Table 5-16	Field bolts (20 bolts per ton - 7/8" dia x 3" avg length)		
		$0.40 x 20/20* = $ 0.40		
	Table 5-24	Field paint $3.50 per ton/20* = 0.18		
		Materials per 100# = $16.08		
		Materials per ton =		
		20 x 16.08 =$321.60		
		Unloading & handling at job site per ton = $ 0.50		
	Table 5-14	Erection - 6 hr per ton @		
	See text	$8.25 per hr = $49.50		
	Table 5-15	Bolting - 20 bolts per ton @		
	See text	6 hr per 100 bolts @ 8.46		
		per hr = $10.15		
	Table 5-24	Painting per ton = $10.30		
		Labor " " = $70.45		
	See equipment estimate	Equipment cost per ton $3571/208.1 tons $17.15		
543	Steel beams	W10 Beams & details	9.1	tons
	Table 5-1	Base cost fob mill per 100# = $ 8.10		
	" 5-2	Quantity extra " " = 0.00		
	" 5-3	Mill extra " " = 1.30		
		Freight to fabricator " " = 0.50		
	" 5-12	Steel detailing " " = 2.00		
	" 5-13	Fabrication " " = 3.30		
		Transportation to job " " = 0.35		
		Total " " = $16.55		
		Fab. overhead 5% " " = 0.83		
		Total " $17.38		
		Fab. profit - 5% " " = 0.87		
		Fab. steel, job site " " $17.25		

Estimate by _____ Date _____ Ckd. by _____ Date _____

Item No.	Identity & Cost Source	Computation of Unit Costs	Total	Units
	Table 5-16	Field bolts (25 bolts per ton 7/8" dia. x 3" avg length) $0.40 x 25/20* = $ 0.50		
	Table 5-24	Field paint $3.75 per ton/20* = 0.19		
		Material per 100# $17.94		
		Material per ton *20 x 17.94 =$358.80		
		Unloading and handling at job site per ton = $ 0.50		
	Table 5-14	Erection - 7 hr " " @		
	See text	$8.25 per hour = 57.75		
	Table 5-15	Bolting - 25 bolts per ton @		
	See text	6 hr per 100 bolts @ $8.46 per hr = 10.15		
	Table 5-24	Painting per ton = 11.30		
		Labor " " = $79.70		
	See equipment estimate	Equipment cost per ton $3571/208.1 tons = $17.15		
544	Base plates	6558#/2000# =	3.3	tons
	Table 5-7	Base cost fob mill per 100# = $ 7.40		
	Table 5-8	Quantity extra: [(6.12 x $2) + (25.76 + 33.70) x 0.70] ÷ 65.58 avg cost per 100# = 0.82		
	Table 5-9	Size extra: [(6.12 x 1.00) + (25.76 + 33.70) 1.90] ÷ 65.58 avg cost per 100# = 1.83		
		Freight to fabricator per 100# = 0.50		
	Table 5-12	Steel detailing per 100# = 1.00		
	Table 5-13	Fabrication " " = 2.00		
		Transportation to job " " = 0.35		
		Total " " $14.65		
		Fab. overhead - 5%" " = 0.73		
		Total $15.38		
		Fab. profit - 5% " " = 0.77		
		Materials " " = $15.40		
		Materials per ton = *20 x $15.40 =$308.00		
		Unloading & handling at job site per ton = $ 0.50		
	Table 5-14	Erection - 3 hr per ton @		
	See text	$8.25 per hr = $24.75		
		Labor per ton $25.25		

SUMMARIES & UNIT COSTS – Example 5-1 (continued)

Estimate by _____ Date _____ Ckd. by _____ Date _____

Item No.	Identity & Cost Source	Computation of Unit Costs	Total	Units
	See equipment estimate	Equipment cost per ton $3571/208.1 tons = $17.15 *Number of 100-lb units in one ton = 2000/100 = 20		

DIRECT COSTS – Example 5-1

Estimate by _____ Date _____ Ckd. by _____ Date _____

Item No.	Identity & Location	Quantity		Unit Cost Each			Total Cost Each			Total Cost
		No.	Unit	Equip.	Mat'l.	Labor	Equip.	Mat'l.	Labor	
540	Steel cols.	52.0	Tons	$ 17.15	$ 352.00	$ 69.45	$ 892.	$18,304.	$ 3,611.	$22,807.
541	Steel bms.	42.5	Tons	17.15	287.40	61.20	729.	12,215.	2,601.	15,545.
542	Steel bms.	101.2	Tons	17.15	321.60	70.45	1,736.	32,546.	7,130.	41,412.
543	Steel bms.	9.1	Tons	17.15	358.80	79.70	156.	3,265.	725.	4,146.
544	Base pls.	3.3	Tons	17.15	308.00	25.25	57.	1,016.	83.	1,156.
							$ 3,570.	$67,346.	$14,150.	$85,066.

OVERHEAD & PROFIT—Example 5-1

Estimate by ——————— Date ——————— Ckd. by ——————— Date ———————

Item No.	Class of Expense	Computations of Overhead Expense	Total Cost
1600	Gen. overhead (% direct cost)	1 1/2% of $85,066	$ 1,276
	Job overhead	Assume 30 days for steel work	
1700	Int. on operating capital	10% of $85,066/2 x 0.09 (int.)/12	32
1701	Superintendent's salary	$10,000/12 months/6 jobs	139
1702	Supt. pickup truck - rental		100
	do. operating cost	1000 miles at 10¢ per mile	100
1703	Job Trucks - rental		
	do. operating cost		
	do. wages - driver's		
1704	Lifting equipment	(Included in equipment)	
	do. operating cost	(Included in equipment)	
	do. wages - operators	(Included in labor)	
1705	Job office - rental		50
	do. salaries		
	do. supplies		25
1706	Utilities & connections	1/4 of job = $200/4	50
1707	Social Security	5.85% x $14,150 x 90%*	745
1708	Workmen's Compensation	12% of $14,150	1,698
1709	Pub. Lia. & Prop. Damage	0.2% of $14,150	28
1710	Fed. & State Unemp. Ins.	3.2% x 45% x $14,150 *	204
1711	Patents & royalties		
1712	Barricades		116
1713	Temporary toilets		50
1714	Cut and patch for trades		
1715	Permits		50
1716	Protection adjacent prop.		
1717	Final cleanup		100
	Subtotal of computed overhead		$ 4,501
1718	Contingencies (% of computed overhead) 10% of $4,501		$ 450
	Total overhead		$ 4,951
	Direct cost from Direct Cost sheet		$ 85,066
	Subtotal (Total overhead plus direct cost)		$ 90,017
1800	Profit 10% of $90,017		$ 9,002
	Total cost		$ 99,019
	Performance bond	$99,019 x 1/1000 x 10	$ 990
	Total amount of bid		$ 100,009
		$100,009 ÷ 208.1 tons = $480.58 per ton	

* Assume average wage of $7.00 per hour

BAR JOISTS

Bar joists are in effect light trusses usually with parallel top and bottom chords which are composed of light angle sections or in some cases special patented shapes. The web is generally of the Warren truss type, being round or square bars for the short-span, open-web joints and angles for the long-span variety. In one form the bottom chord rests on the supporting member while in the other or underslung type the top chord rests on the support with the last bay of the bottom chord bent up and welded to the top chord to form a seat. The assembly line method of welded construction makes bar joists very competitive with steel beams.

One advantage of bar joists is that they are relatively light as compared to beams of the same span length and are more easily and quickly erected. Also the open-web feature makes an ideal raceway through which to route the several building services. Joists can be supported on masonry walls or steel beams. They are usually welded to the beams and fastened to the masonry walls with plates and anchors. The joist must be held in place with cross bracing at about 5- to 10-foot centers.

Concrete floors and roofs are constructed using bar joists with ribbed or corrugated metal forms of various designs and shapes with the ribs running perpendicular to the joists. These metal forms are left in place and save the use of wood forms with their supporting beams and shores, which are slower to use and more expensive in the overall picture most of the time. In fire-proof construction metal lath for plaster can be fastened to the bottom chord of the joist or suspended therefrom.

For wood floors joists can be bought with the wood nailers factory fastened to the top chord, thus saving erection time in the field.

Table 5-26 shows some approximate prices for bar joists complete with the required bracing. However, since this product is such a competitive item, the plans and specifications should be sent to several bar joist manufacturers to get the best price for the joists delivered to the job site.

Table 5-27 shows some approximate times for erection.

TABLE 5-26
Approximate Cost of Bar Joists

Span of joists	Cost per ton
6 feet to 50 feet	$235 to $300
50 feet to 100 feet	250 to 310

Note: In small quantities these amounts would be higher.

114   *Chapter Five*

TABLE 5-27

Approximate Erection Times for Bar Joists

Type of construction	Man-hours erecting	Man-hours welding
Simple with large areas	6 - 8	2.5*
Irregular with small areas	10 - 12	3.5*

*For welding cross bracing and ends of joists to beams see section on welding on p. 98 .

The cost of the weld rod can be estimated from information given in Table 5-22 and the cost of the welding machine can be determined from Appendix A.

The cost of cranes to lift the bar joists to the various floors where they can be spread out by hand will be similar to the cost of lifting reinforcing bars per ton and this cost has already been discussed in the text.

The cost of placing metal decking on bar joists for use as formwork for concrete will vary from $0.15 to $0.30 per square foot and the decking itself from $1.00 to $1.50 per square foot depending upon the type of decking used. There are many different types of this material and as in the case of the joists themselves it is an extremely competitive product and the prices vary considerably even for similar designs.

HOMEWORK EXAMPLE 5-1

This example consists of the structural frame for a small office building. The student will make a complete estimate of the structural framework erected in place, plumbed, and painted using local prices for all materials and wages.

Figure 5-2, the drawings for use with Homework Example 5-1, are self-explanatory.

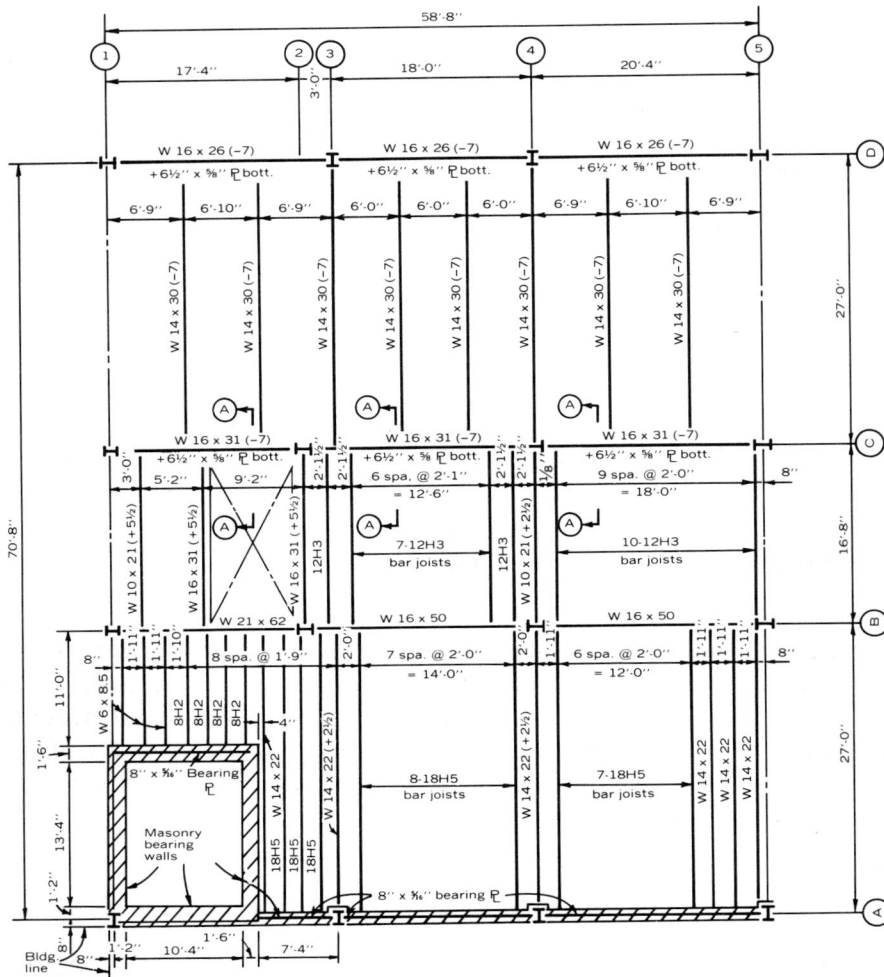

1st FLOOR FRAMING PLAN—TOP OF STEEL EL. 380'-6½''°

°UNLESS NOTED ±x''

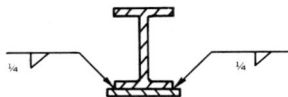

SECTION A-A

FIGURE 5-2

Illustrations for Homework Example 5-1

A.   First-Floor Framing Plan for Homework Example 5-1

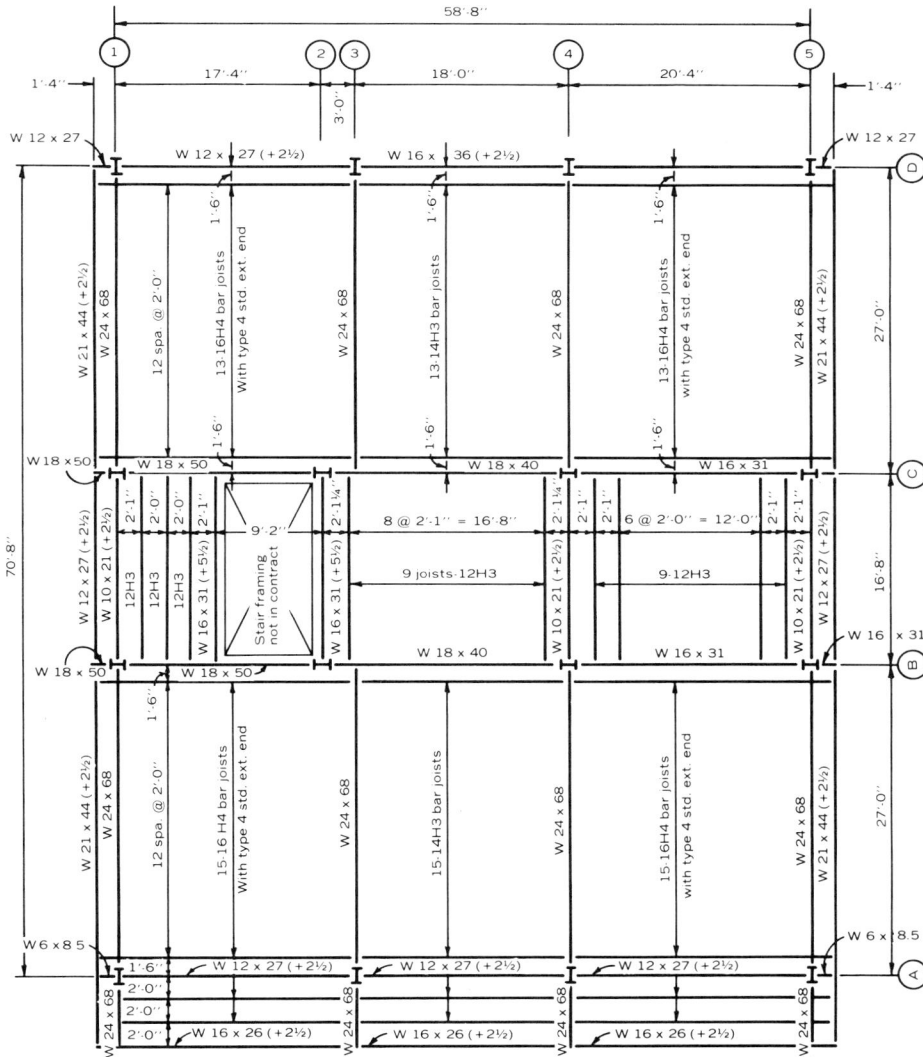

2nd FLOOR FRAMING PLAN—TOP OF STEEL EL. 393'-10½''°

3rd FLOOR FRAMING PLAN—TOP OF STEEL EL. 407'-2½''°

°UNLESS NOTED ±x''

FIGURE 5-2 (continued)

B.   Second- and Third-Floor Framing Plan

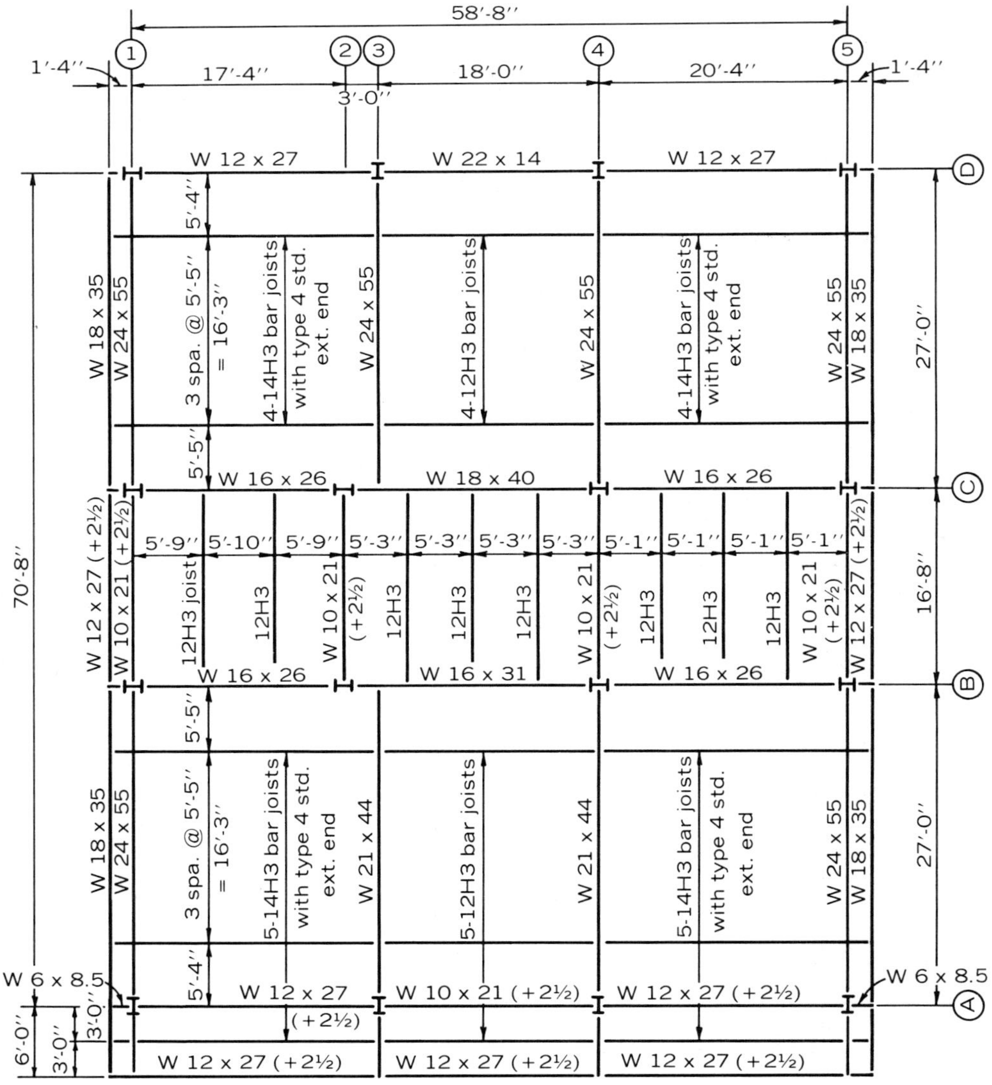

ROOF FRAMING PLAN—TOP OF STEEL EL. 420'-9½''*

*UNLESS NOTED ±x''

FIGURE 5-2 (continued)

C.   Roof Framing Plan

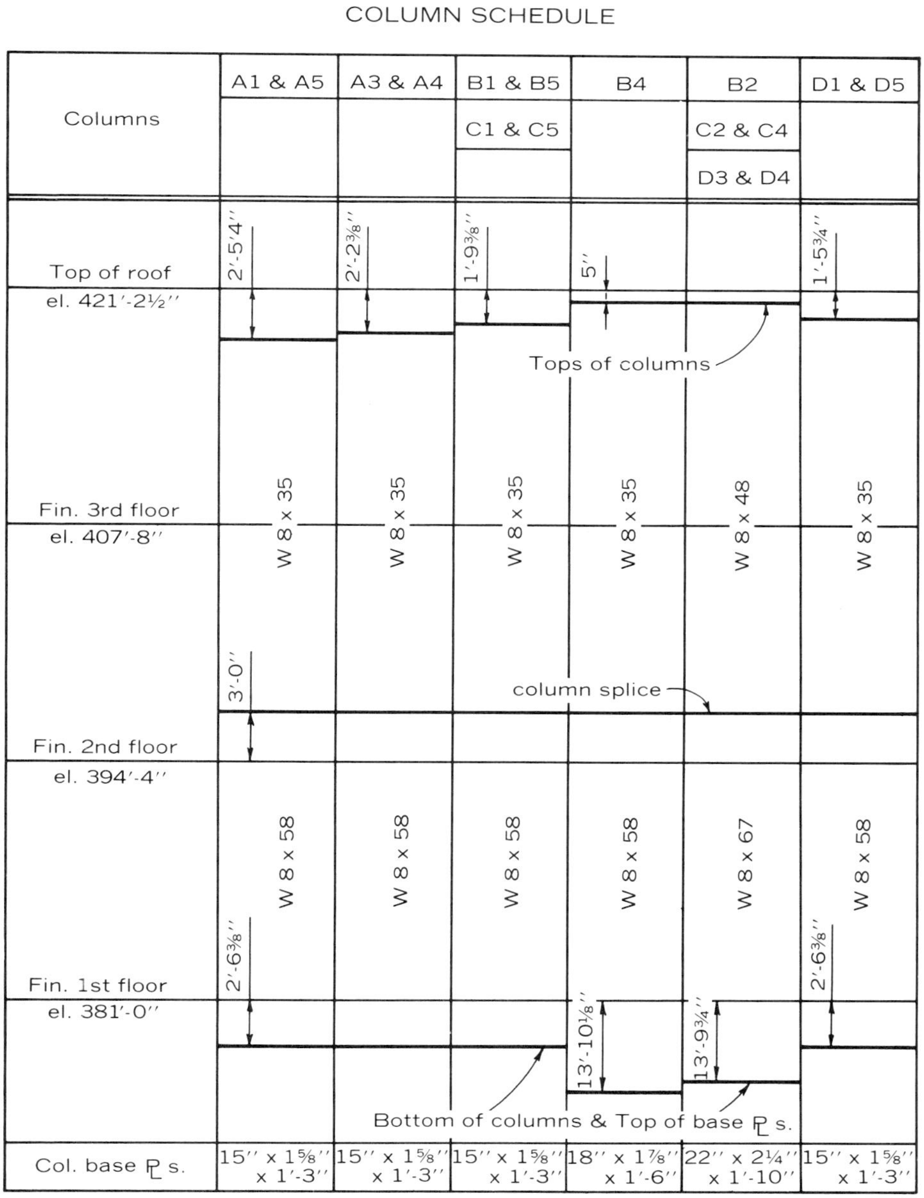

COLUMN SCHEDULE

Columns	A1 & A5	A3 & A4	B1 & B5 / C1 & C5	B4	B2 / C2 & C4 / D3 & D4	D1 & D5
Top of roof el. 421'-2½"	2'-5¼"	2'-2⅜"	1'-9⅜"	5"		1'-5¾"
Fin. 3rd floor el. 407'-8"	W 8 × 35	W 8 × 35	W 8 × 35	W 8 × 35	W 8 × 48	W 8 × 35
Fin. 2nd floor el. 394'-4"	3'-0"					
Fin. 1st floor el. 381'-0"	W 8 × 58 (2'-6⅜")	W 8 × 58	W 8 × 58	W 8 × 58 (13'-10⅛")	W 8 × 67 (13'-9¾")	W 8 × 58 (2'-6⅜")
Col. base ℄ s.	15" × 1⅝" x 1'-3"	15" × 1⅝" x 1'-3"	15" × 1⅝" x 1'-3"	18" × 1⅞" x 1'-6"	22" × 2¼" x 1'-10"	15" × 1⅝" x 1'-3"

Tops of columns

column splice

Bottom of columns & Top of base ℄ s.

FIGURE 5-2 (continued)

D. Column Schedule

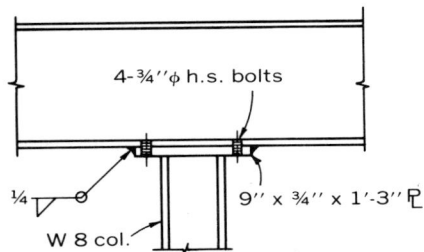

TYPICAL SECTION
WHERE ROOF B'MS. CROSS COLS.

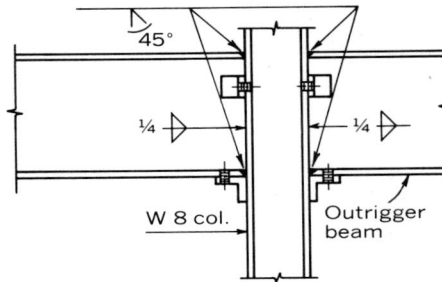

4-¾''φ h.s. bolts

9'' x ¾'' x 1'-3'' ℙ

W 8 col.

TYPICAL SECTION
AT OUTRIGGER BEAMS

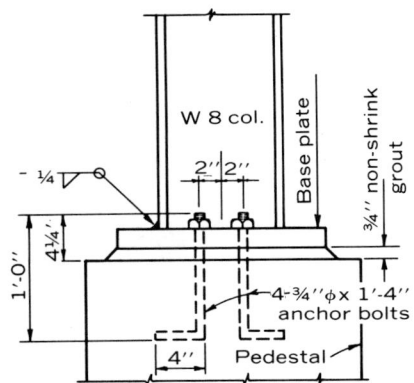

45°

¼

¼

W 8 col.

Outrigger
beam

TYPICAL ANCHOR
BOLT SETTING

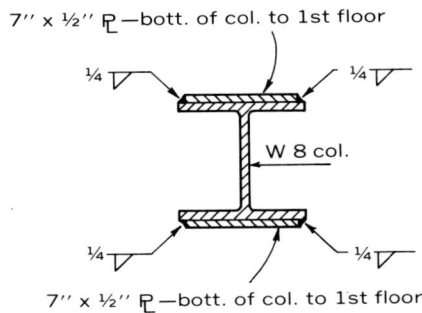

W 8 col.

Base plate

¾'' non-shrink grout

2'' 2''

4¼''

1'-0''

4''

Pedestal

4-¾''φ x 1'-4''
anchor bolts

TYPICAL AT COLS
B2
C3 & C4
D3 & D4

7'' x ½'' ℙ —bott. of col. to 1st floor

¼

¼

W 8 col.

¼

¼

7'' x ½'' ℙ —bott. of col. to 1st floor

SPECIFICATIONS

All steel A36

All connections field-bolted with
¾''φ h.s. ASTM-A325 bolts unless noted.

All details in accordance with AISC
standard requirements.

FIGURE 5-2 (continued)

E.   Typical Sections

120   *Chapter Five*

CHAPTER SIX

Masonry

KINDS OF MASONRY

The most commonly used masonry units are brick, concrete block, tile, and stone. Walls are laid up of each of these materials separately or in combination, such as face brick backed up by common brick or concrete block.

Brick Masonry

The Number of Brick Required per Sq Ft of Wall. One of the best ways to count the number of brick in a wall is to establish the number of brick in a square foot and then multiply this number by the total square feet required. This will give the number of brick for a wall of one thickness (or "wythe"), which for a standard size brick is 4". The quantity of brick in a wall of 8" or 2 wythes would be twice this number. For a wall of 12" or 3 wythes the quantity would be three times the number in the 4" wall and so on.

The size of the mortar joint has quite a lot of influence on the number of brick required. The standard brick is 2 1/4" × 3 3/4" × 8" with the 2 1/4" × 8" dimension visible in the wall. With a 1/4" joint the number of brick in a square foot would be 144/(2 1/2 × 8 1/4) = 7. In an 8" wall the number would be 2 × 7 = 14. In a 12" wall the number would be 3 × 7 = 21, and so on. Most walls are of standard brick with 1/4" mortar joints. For the different size joints using standard brick the number of brick per square foot in a 4" thick wall will be:

Joint sizes—	1/4"	5/16"	3/8"	1/2"	5/8"	3/4"
Brick/sq ft—	7	6.75	6.55	6.15	5.80	5.50

If the brick are laid in a bond (pattern), such as running bond, where no half brick are used, and if no headers are required, then the quantities suggested will be adequate. However, if header courses are required, then the number of brick per square foot will have to be increased. After the required number of brick is determined the amount should be increased by 3 to 5% to allow for waste and breakage.

Sometimes it is more convenient to compute the number of brick by the cubical content. An average number of brick per cubic foot of masonry for the standard size of brick is:

Joint sizes—	1/4″	5/16″	3/8″	1/2″
Brick/cu ft—	21.4	20.5	19.5	17.5

In estimating the number of brick in an irregular volume like a chimney, it is better to lay out a section through the chimney using the size of the brick specified and the desired size of joint, and then count the number of brick in a layer. The number of layers will be the height of the chimney in inches divided by the depth of the brick plus the size of the joint. Add 10% breakage for chimney.

Estimating Brick Quantities. In estimating the number of brick required in a wall subtract the areas for openings from the gross area, compute the number of brick indicated for this area, and then add the extra brick needed for trim. In the case of face brick backed up by common brick or block add 1.5 face brick per linear foot of openings around windows and around doors except at bottom of doors. In the case of brick veneer add 5 brick per foot at tops of windows and doors for soldier course and 5 brick per foot at bottoms of windows for rowlock course. Brick are always estimated in units of 1,000.

Estimating Labor for Laying Brick. The tables in this chapter show a variation in the time required for laying brick. The lower times are for simple work, such as running bond with few openings. The longer times are caused by more intricate bond patterns and a greater number of openings.

It is usually customary to figure one helper with each mason. The masons lay the brick while the helpers keep them supplied with brick and mortar, set up and move the scaffolding, and operate the mortar mixer. So the labor price per hour for laying brick will include the hourly wage of the mason plus the hourly wage of helper. The tables shown include the labor required for pointing and cleaning both brick and block masonry.

Concrete Block Masonry

Types of Concrete Block Masonry. There are two principal types of concrete blocks: lightweight block made by using a lightweight aggregate, such as solite, in the concrete mix; and the block made by using the regular coarse aggregate in the mix. This latter gives a much heavier product which costs more to lay because it is a great deal more difficult to handle.

Both of these blocks are made in regular and in thickwall where the cavities are smaller, thus making the walls thicker and the resulting block heavier and with greater load-bearing potential than the regular-weight block. They are both also made solid without openings.

Practically all the block now used are lighweight because of the savings in the cost of shipping and laying as well as in the reduction of dead load, thus effecting further savings in the supporting members and foundations. Where load-bearing walls of high capacity are required the thicker wall block is used while below grade the solid block or heavyweight aggregate concrete block is often specified.

In our tables and examples we will consider only the regular lightweight block.

The Number of Block Required per Sq Ft of Wall. The nominal size of block is $8'' \times 16'' \times$ the different thicknesses, which are, usually, $2''$, $4''$, $6''$, $8''$, $10''$, and $12''$. These are the dimensions center to center of the mortar joints, if the joint being used is $3/8''$ thick, thus making neat size $7\ 5/8'' \times 15\ 5/8'' \times 1\ 5/8''$, etc.

Block are generally laid in running bond or stacked or in some similar pattern with the $8'' \times 16''$ face showing. With a $3/8''$ joint the number of block required per square foot of wall would be $144/8 \times 16 = 1.125$ or $1\ 1/8$ block. For different size joints the procedure would be the same, giving approximately $1\ 1/7$ block for $1/4''$ joints and 1.1 block for $1/2''$ joints. To these quantities should be added 2 to 4% for waste and breakage.

Estimating Block Quantities. In estimating the number of block required in a wall laid up in running bond subtract the area for openings from the overall area to get the net area to be covered by the block.

Next compute the number of lintel block, jamb block, half-length jamb block, steel sash block, half-length steel sash block, half-length regular, and other special block called for on plans.

To obtain the actual number of regular block required compute the number of block needed to fill up the net area and then subtract the number of special block which will be used in the wall.

In running bond vertical lines through the joints will cut every other course of block at its midpoint of length. Therefore half-length block are needed at every other course to end the wall in a flush line.

Estimating Labor for Laying Block. The tables in this chapter show a variation in the time for laying block. As in the case of brick the complexity of the work will control the cost of laying the block.

One helper is usually allowed for each mason as an average. The masons lay the block and the helpers keep them supplied with block and mortar, set up and move the scaffolding, and operate the mortar mixer. Thus the labor price per hour for laying block will include the hourly wage of the mason plus the hourly wage of the helper.

In both brick and block work one of the masons usually acts as the leader or straw boss and so it is not necessary to charge the time of a foreman to this work.

Brick Bonds

Four of the most commonly used brick bonds are shown in Figure 6-1.

In common bond every sixth course is a header course which requires twice as many brick as a stretcher course. Therefore, when the number of

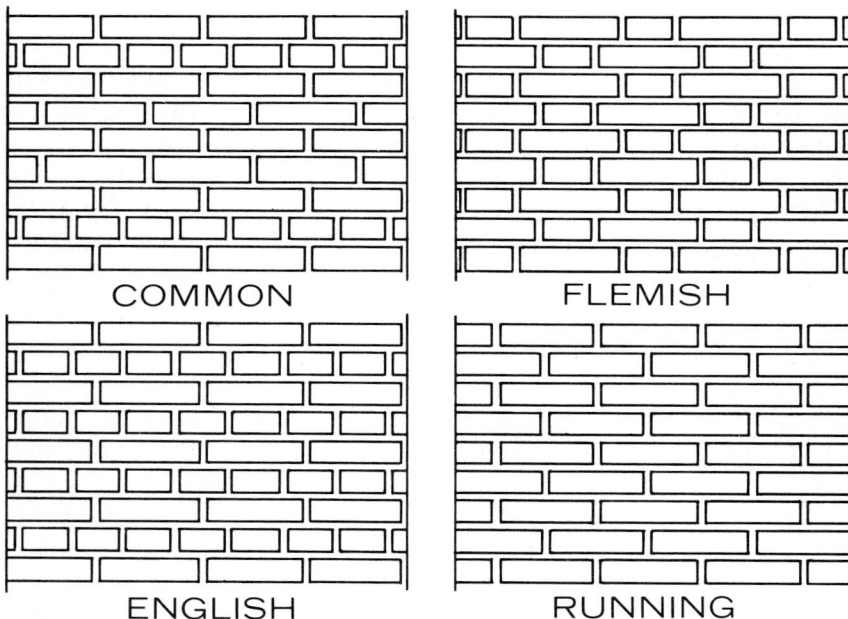

FIGURE 6-1
Four Standard Brick Bonds

brick required is computed as so many per square foot it must be increased by one-sixth to allow for the extra brick in the header courses.

In brick veneer work such as dwellings the brick are usually laid up in running bond, which is the same as common bond without the extra brick required for the headers.

In English bond every other course is a header course. So when the number of brick is figured by the square foot method that number must be increased by one-half. In Flemish bond this increase will be one-third.

Material and Labor Quantities

Tables 6-1 to 6-4 show approximate quantities and laying time for brick and block work. (Metal ties are used to anchor brick masonry to the backup material against which it is laid. Allow one tie per square foot of wall. The cost will be approximately $0.02 per tie delivered to the job. The mason will ordinarily put up the ties as he goes so no allowance need be made for labor.)

TABLE 6-1

A. Approximate Quantities of Mortar Required
per 1,000 Brick* (For full joints in common bond)

Joints	1/4″	5/16″	3/8″	7/16″	1/2″
Cu yd of mortar	0.33	0.36	0.42	0.52	0.55

B. Materials Required for One Cubic Yard of
Mortar for Brick Masonry

Mix by volume cement-lime-sand			Sacks of cement	Lb of lime	Cu yd of sand
1	1/4	3	9.0	91	1
1	1/2	4	6.8	135	1
1	1	5	5.4	211	1
1	1	6	4.5	181	1
1	2	6	4.5	360	1
1	2	9	3.0	242	1

*The mortar required varies for different size walls. This table was made for a 12-in. wall. For a 4-in. wall use 93% of the above; for an 8-in. wall use 96%; and for a 16-in. wall use 103%.

TABLE 6-2
Average Time For Laying Brick in Common Bond
(Including pointing and cleaning)

Type of work	Mason hours per 1,000 brick*
Common brick walls (finished one side)	5 - 12
Common brick walls (finished two sides)	6 - 14
Face brick walls	10 - 20
Fire brick walls	18 - 32

Notes: For other than common bond increase mortar and labor hours 10 to 15%.

Consider running bond same as common for mortar and labor hours.

For house chimneys double the manhours in this table.

*Add same number of hours for helper.

TABLE 6-3
Approximate Quantities of Mortar Required for
100 Standard Size Concrete Block

Joints	Approximate cu yd of mortar
3/8"	Varies from 0.16 to 0.32 due to size and net cross-sectional area of block

Materials Required for One Cubic Yard of Mortar
for Block Masonry (Based on the ratio of 10 lb of
lime to one bag of cement)

Mix by volume cement - sand		Sacks of cement	Lb of lime	Cu yd of sand
1	2	13.5	135	1
1	3	9.	90	1
1	4	6.8	70	1

TABLE 6-4
Average Time for Laying Lightweight Concrete
Block* (Including pointing and cleaning)

Block size	Mason hours per 100 block†
2 × 8 × 16	3.0 - 3.5
4 × 8 × 16	3.5 - 4.0
6 × 8 × 16	4.0 - 4.5
8 × 8 × 16	4.7 - 5.2
10 × 8 × 16	6.0 - 6.5
12 × 8 × 16	7.0 - 7.5

*Add 10% for heavyweight aggregate block.
†Add same number of hours for helper.

Material Costs

Tables 6-5 through 6-9 show approximate costs of materials for brick and block work as of 1971.

TABLE 6-5
Approximate Average Costs for Mortar Materials

Lime	$2.00 - $3.00	per 100 lb
Sand	3.50 - 6.00	per cu yd
Cement	1.25 - 1.75	per bag

Note: One cubic yard of sand weighs approximately 1 1/3 tons.

TABLE 6-6
Approximate Price per 1,000 for Standard Size Brick

Common brick	$ 35 - $ 70
Face brick	45 - 85
Fire brick	200 - 270

TABLE 6-7
Approximate National Average Price per 100 for
Concrete Block (Made from lightweight aggregate*)

2 X 8 X 16	Solid	$ 22.00
3 X 8 X 16	Regular	22.00
4 X 8 X 16	Solid	$ 34.00
4 X 8 X 16	Regular	22.00
4 X 8 X 8	Half	16.50
4 X 8 X 16	Corner	35.00
4 X 8 X 16	Bond beam lintel block	34.00
6 X 8 X 16	Two hour	$ 33.00
6 X 8 X 16	75% solid (4 hour)	35.00
6 X 8 X 16	Regular	30.00
6 X 8 X 8	Half	23.00
6 X 8 X 16	Corner	37.50
6 X 8 X 16	Steel sash	32.00
6 X 8 X 16	Bond beam lintel block	36.00
8 X 8 X 16	75% solid	$ 54.00
8 X 8 X 16	Solid	65.00
8 X 8 X 16	Regular	35.00
8 X 8 X 8	Half	23.00
8 X 8 X 16	Double corner	37.50
8 X 8 X 16	Wood sash	37.50
8 X 8 X 16	Wood sash half	24.00
8 X 8 X 16	Bond beam lintel block	42.00
8 X 8 X 16	Chimney block	55.00
10 X 8 X 16	Regular	$ 46.00
10 X 8 X 8	Half	30.00
10 X 8 X 16	Corner	48.50
10 X 8 X 16	Steel sash	48.50
12 X 8 X 16	75% solid	$ 70.50
12 X 8 X 16	Four hour block	55.00
12 X 8 X 16	Regular	48.50
12 X 8 X 8	Half	33.00
12 X 8 X 16	Corner	48.50
12 X 8 X 16	Bond beam lintel block	56.00
16 X 8 X 19	Pilaster block (2 piece)	$110.00
2 1/4 X 4 X 8	Concrete brick per 1,000	$ 60.50

*For heavyweight aggregate cost will be approximately
10% less.

TABLE 6-8
Approximate Price of Precast Concrete Lintels per
Linear Foot (Made from lightweight aggregate)

4 × 8 lintels	$1.60
6 × 8 lintels	1.80
8 × 8 lintels	2.10
12 × 8 lintels	2.50

TABLE 6-9
Some Approximate Costs of Equipment
for Laying Brick

1.	Hand tools and misc. equip.	allow	$1.00 per 1,000 brick
2.	Mortar boxes, mixer and hoist	allow	$1.00 to $3.00 per 1,000 brick
3.	Scaffolding	allow	$2.00 to $6.00 per 1,000 brick

Note: The equipment cost to lay 100 block will be roughly
equal to the equipment cost per 1,000 brick.

The cost of common and face brick will be influenced by the required texture, finish, size, and shape of the brick. However, none of these will affect the cost as much as the requirements of a modular size permitting no variations in dimensions such as the number of brick required to lay up a brick and block wall in which 3 brick layers with the required joint size will exactly match the nominal height of the block.

Equipment

Bricklaying Equipment. The equipment required for laying brick will consist of such things as: mortar boxes, hand tools, mortar mixers, manual or power hoists, shovels, scaffolding, etc.

General Use in Building Work. The use of stone masonry for buildings is declining in favor of materials such as glass, aluminum, and cast stone panels. The cast stone panels are made by casting stone of various sizes and hues into a concrete base in varying patterns. The product is extremely flexible and can be made to achieve almost any desired size and shape. These panels will be discussed in chapter 12.

The use of stone in exterior veneers and interior trims is now generally confined to government buildings, banks, large churches, cathedrals, and other monumental buildings of the traditional classic types. In certain foreign countries such as Greece and Italy the marbles and granites are the cheapest building materials available. However this material is usually too costly for the ordinary building in this country.

The cost of stone facing will depend on the type of the stone, its finish, and its thickness. This price can vary from about $5.00 to $20.00 per square foot.

The Use of Stone in Small Building Work. Very often orchard stone, which is a rough unsquared stone, is used for the facing of small buildings and residences. Random size, low-priced stone, approximately four inches in thickness, also falls in this category.

Cost for a four-inch thick wall of this type may be computed as follows:
  Material
        Cost of stone—will vary from about $0.50 to $2.00
        per sq ft
        Mortar reqd. per 100 sq ft of wall—0.2 to 0.3 cu yd
  Labor
        Crew of one mason and three helpers
        can lay 100 sq ft of wall in 3 to 9 hours.
  Equipment
        For miscellaneous tools, mortar boxes, mixer,
        hoists, and scaffolds, allow $3.00 to $7.00 per 100 sq ft.

Extra Cost Factors

Height of Work. A scaffold floor is required about every 4 feet of height to enable a mason to reach his work. As the work goes higher there is an additional cost due to having to lift the masonry and mortar to the new height.

This extra work can be accomplished by allowing ½ helper per mason for every added scaffold (4-ft interval). Another method is to provide a portable lift with operator that will place the masonry and mortar on the scaffold. One such outfit can service all of the masons on a medium-sized job.

In some areas the mason's hourly rate increases with added height. This is especially true in chimneys where scaffolding and the required elevator to lift materials also add considerably to the cost of masonry in place.

Wall Openings. Walls which are interrupted with many openings increase labor cost due to extra work required to plumb the jambs. These extra cost factors will not be considered in Example 6-1 in the interest of keeping the estimate simple; however, they should be taken into account in larger jobs.

EXAMPLE 6-1

Estimate the cost of curtain and partition walls of brick and block required to cover the steel frame in Figure 5-1. Assume that proper steel plates are welded to tops of spandrel beams to carry the brick.

The exterior walls will be 4 in. of face brick laid in running bond backed up with 4 in. of lightweight concrete block. This wall will be built on top of the grade beams, which stop 4 in. below the finished first floor. The face brick will continue unbroken to top of parapet wall; however, the block will be interrupted by the beams and slabs. Brick joints will be 3/8″+ and block joints 3/8″. On the first floor there will be 4 large doors, each 8′ × 8′, and 4 small doors, each 3′ × 7′.

There will be no partition walls in the basement. On each of the other 3 floors there will be 3 partition walls in the north-south direction, all of 4″-lightweight concrete block laid with 3/8″ joints in running bond. There will be doors 4′ × 7′ on each floor in partition walls.

Floors will be of 3-in. concrete and the roof of 2-in. concrete laid on tops of beams. All lintels will be of steel and will be ignored in this estimate as will all flashing. The metal ties used to hold the block to the beams and column will be welded in the shop and estimated with the steel frame.

Allow one metal tie per square foot to secure the brick to the block at a cost of $0.02 each for material. The cost of installation is included in the brick mason's wage cost.

There will be an 8-in. brick parapet extending 3 ft above roof line. The parapet will be made of face brick backed up by common brick, both laid up in common bond with 3/8″ mortar joints.

Since this is a storage warehouse there will be no windows.

QUANTITY TAKEOFF—Example 6-1

Estimate by _____ Date _____ Ckd. by _____ Date _____

Item	Identity	Location	Quantity Computations	Total	Units
600	Half 4" Conc block L.W.	E-W Outside walls	12 (34.58' ÷ .667') x 1.03 =	641	
	"	N-S Outside walls	6 (32.58' ÷ .667') x 1.03 =	302	
	"	Outside doors	[(4 doors x 8') + (4 doors x 7') ÷ .667'] x 1.03 =	93	
	"	Inside doors	3 Floors (17 doors x 7') ÷ .667' x 1.03 =	552	
	"	E-W Inside walls	12 (34.58' ÷ .667') x 1.03 =	641	
	"	N-S Inside walls	9 (32.58' ÷ .667') x 1.03 =	453	
			TOTAL:	2,682	

NOTE: When block are laid up in running bond one half block will be required for each course for every length of wall.

In order to arrive at the number of regular block needed, compute the block required for area and subtract half the number of half block required.

Estimate by _____ Date _____ Ckd. by _____ Date _____

Item	Identity	Location	Quantity Computations	Total	Units
601	4" Light weight conc. block	N-S ext. walls	$2[(32.58' \times 57') - (1' \times 8' \times 8') - (1' \times 7' \times 4')] \times 1.125 \times 1.03$ (Breakage) =	4,091	Each
	"	E-W ext. walls	$2[(34.58' \times 120') - (1' \times 8' \times 8') - (1' \times 7' \times 4')] \times 1.125 \times 1.03 =$	9,404	
	"	N-S part. 1st floor	$3[(9.75' \times 57') - (3' \times 4' \times 7')] \times 1.125 \times 1.03 =$	1,640	
	"	E-W part. 1st floor	$2[(10.58' \times 120') - (4' \times 4' \times 7')] \times 1.125 \times 1.03 =$	2,681	
	"	N-S part. 2nd floor	$3[(9.75' \times 57') - (3' \times 4' \times 7')] \times 1.125 \times 1.03 =$	1,640	
	"	E-W part. 2nd floor	$2[(10.58' \times 120') - (4' \times 4' \times 7')] \times 1.125 \times 1.03 =$	2,681	
	"	N-S part. 3rd floor	$3[(11.33' \times 57') - (3' \times 4' \times 7')] \times 1.125 \times 1.03 =$	1,953	
	"	E-W part. 3rd floor	$2[(11.83' \times 120') - (4' \times 4' \times 7')] \times 1.125 \times 1.03 =$	3,030	
			Total block spaces Total half block ÷ 2 =	27,120 -1,340	
			Total regular block Use	25,780 25,800	
610	Standard face brick	E-W ext. walls	$2[(41' \times 121.67') - (1' \times 8' \times 8') - (1' \times 7' \times 4')] \times 6.55 \times 1.04$ (Breakage) =	66,709	
	"	N-S ext. wall	$2[(41' \times 61') - (1' \times 8' \times 8') - (1' \times 7' \times 4')] \times 6.55 \times 1.04 =$	32,821	
	" "	Top of doors Door jambs	$5[(4' \times 8') + (4' \times 3')] \times 1.04 =$ $1.5[(4' \times 8' \times 2') + (4' \times 7' \times 2')] \times 1.04 =$	229 187	
	"	8" parapet wall x 3'	$6.55(121.67' + 61') 2 \times 3 \times 1\,1/6 \times 1.04$	8,707	
			Total face brick	108,653	

QUANTITY TAKEOFF—Example 6-1 (continued)

Estimate by _____ Date _____ Ckd. by _____ Date _____

Item	Identity	Location	Quantity Computations	Total	Units
611	Standard common brk.	8" parapet wall x 3'	$6.55(121.' + 60.33') \; 2 \times 3 \times 5/6 \times 1.04 =$	6,236	
612	Steel wall ties	N-S ext. walls	$2[(32.58' \times 57') - (1' \times 8' \times 8') - (1' \times 7' \times 4')] =$	3,526	
		E-W ext. walls	$2[(34.58' \times 120') - (1' \times 8' \times 8') - (1' \times 7' \times 4')] =$	8,116	
			Total wall ties	11,642	

SUMMARIES & UNIT COSTS—Example 6-1

Estimate by _____ Date _____ Ckd. by _____ Date _____

Item No.	Identity & Cost Source	Computation of Unit Costs	Total	Units
600	4" Lightweight conc. block	Including 3% breakage = 25,780 order-	258	100
	3/8" joints 1-3 mortar mix	Cost per 100:		
x xx	Mortar/100 block Mat'ls/cu yd of Mortar See Table 6-3	Block x xx = $22.00 Cement .30(9. x $1.60/sack) = 4.32 Lime .30(90/2000 x $50/ton) = .68 Sand .30(1. x 4.00/cu yd) = 1.20 Material cost $28.20		
xxx	Hr to lay 100 block See Table 6-4	Mason & helper = $8.40 + $6.20 = $14.60 Labor = 3.75 hr/100 @ $14.60 = $54.75 xxx		
	See Table 6-9	Mortar boxes, mixer 2.00 Small tools & misc equip. 1.00 Scaffold 2.00 Misc 2.00 Equipment $ 7.00		
601	Half - 4" Light-weight conc. block	Including 3% breakage = 2682 order-	27	100
		Block = 16.50 (Half of cement, lime,and sand ⎰ 2.16 quantities shown above) ⎨ .34 ⎱ .60 Material cost $19.60		
		Labor (See above) $54.75		
		Equipment (See above) $ 7.00		

Estimate by _____ Date _____ Ckd. by _____ Date _____

Item No.	Identity & Cost Source	Computation of Unit Costs	Total	Units
610	Standard face brick	Including 4% breakage = 108,653 order =	108.7	1,000
	3/8 joints 1-1-5 mortar mix	Cost per 1,000:		
x xx	Mortar/1,000 brk Mat'ls/cu yd of mortar See Table 6-1	Brick x xx = $ 70.00 Cement 0.42 (5.4 x $1.60/sack) = 3.63 Lime 0.42 (211/2000 x $50/ton) = 2.22 Sand 0.42 (1 x $4.00/cu yd) = 1.68 Material cost $ 77.53		
xxx	Hours to lay 1,000 brick See Table 6-2	Mason & helper = $8.40 + $6.20 = $14.60 Labor = 12 hr /1000 @ $14.60 = $175.20 xxx		
	See Table 6-9	Mortar boxes, mixer & hoist = 2.00 Small tools & misc equip. = 1.00 Scaffold = 2.00 Misc = 2.00 Equipment $ 7.00		
611	Standard common brick	Including 4% breakage = 6,236 order =	6.3	1,000

		Brick = $ 55.00
		From above 3.63
		     " 2.22
		     " 1.68
		Material cost $ 62.53		

		Labor 8 hr /1,000 @ $14.60 = $116.80		
		Equipment (From above) = $ 7.00		
612	Metal ties	11,642 @ 2¢ each (Total) = $233.00		

DIRECT COSTS—Example 6-1

Estimate by _____ Date _____ Ckd. by _____ Date _____

Item No.	Identity & Location	Quantity		Unit Cost Each			Total Cost Each			Total Cost
		No.	Unit	Equip.	Mat'l.	Labor	Equip.	Mat'l.	Labor	
600	4" Conc. blk.	258	100	$ 7.00	$ 28.20	$ 54.75	$1,806.	$ 7,276.	$14,126.	$23,208.
601	4" Half blk.	27	100	7.00	19.60	54.75	189.	529.	1,478.	2,196.
610	Face brick	108.7	1,000	7.00	77.53	175.20	761.	8,428.	19,044.	28,233.
611	Com. brick	6.3	1,000	7.00	62.53	116.80	44.	394.	736.	1,174.
612	Metal ties								233.	233.
							$2,800.	$16,627.	$35,384.	$54,811

OVERHEAD & PROFIT—Example 6-1

Estimate by _____ Date _____ Ckd. by _____ Date _____

Item No.	Class of Expense	Computations of Overhead Expense	Total Cost
1600	Gen. overhead (% direct cost)	2% of $54,811	$ 1,370
	Job overhead	(Allow 2 months for masonry work)	
1700	Int. on operating capital	10% of $54,811/2 x 0.09 x 2/12 (months)	41
1701	Superintendent's salary	12,000/12 x 2 (1/3 of time to masonry)	667
1702	Supt. pickup truck – rental		200
	do. operating cost	3,000 miles @ 10¢ per mile	300
1703	Job trucks – rental		
	do. operating cost		
	do. wages – driver's		
1704	Lifting equipment	Gasoline hoist & tower	300
	do. operating cost		50
	do. wages – operator's	(1/2 of operators time @ $7.95/hr)	1,367
1705	Job office – rental		50
	do. salaries		
	do. supplies		35
1706	Utilities & connections	1/4 is charged to masonry = $400 ÷ 4	100
1707	Social Security	5.85% x $35,384 x 90%*	1863
1708	Workmen's Compensation	3% x $35,384	1,062
1709	Pub. Lia. & Prop. Damage	0.2% x $35,384	71
1710	Fed. & State Unemp. Ins.	3.2% x 45% x $35,384*	510
1711	Patents & royalties		
1712	Barricades		100
1713	Temporary toilets		50
1714	Cut and patch for trades		284
1715	Permits		50
1716	Protection adjacent prop.		
1717	Final cleanup		100
	Subtotal of computed overhead		$ 7,784
1718	Contingencies (% of computed overhead) 10% of $7,784		$ 778
	Total overhead		$ 8,562
	Direct cost from Direct Cost sheet		$ 54,811
	Subtotal (Total overhead plus direct cost)		$ 63,373
	Profit 12% of $63,373		$ 7,605
	Total cost		$ 70,978
	Performance bond $70,978 x 1/1000 x $10		$ 710
	Total amount of bid		$ 71,688
	*Assume average wage of $7.00/hr		

Estimate the cost of curtain and partition walls of bricks and block required to cover the second and third floors of the steel frame shown in Figure 5-2.

The exterior walls will be of face brick backed up by 4-in. lightweight concrete block both laid up in running bond with 3/8″ joints. Use wire mesh joint reinforcement at every other block (or six-brick) joints to tie the brick to the block. Ignore the metal ties to hold the block to the beams and columns since they will be welded in the shop and included in the cost of the structural steel framing. Assume that exterior walls will be carried by the spandrel beams that run from center to center of exterior columns and that these beams will have steel plates welded to their top flanges to carry the brick.

There will be 6-in. lightweight concrete block partition walls at column lines B and C and 4-in. lightweight concrete block walls at column lines 3 and 4 between column lines A to B and C to D. Partition wall will be laid up in running bond with 3/8″ joints.

There will be 21 windows 4′ × 7′ in exterior walls on each floor and 12 doors 3′ × 7′ in partitions on each floor. Make no allowance for lintels.

CHAPTER SEVEN

Wood Construction

CLASSES OF CONSTRUCTION

Wood construction can be divided roughly into two classes:
1. Heavy construction
   - a.   Mill buildings
   - b.   Towers
   - c.   Trusses
   - d.   Wharves
   - e.   Warehouses
   - f.   Fertilizer plants, etc.
2. Light construction, which is principally housing, is usually divided into two parts for the purpose of pricing:
   - a.   Rough carpentry
   - b.   Finish carpentry

This text will confine itself to the problems of light construction.

MATERIALS

The cost of framing lumber will vary from less than $100 to more than $200 per thousand board feet depending upon the thickness and the width of material, grade of lumber, and location of the work. Some average prices are:

2″-thick material through 2″ × 8″	— $150 mfbm
2″ × 10″ and 2″ × 12″	— 160 ″
3″ thick material begins at about	— 185 ″
4″ thick material begins at about	— 200 ″
6″ thick material begins at about	— 215 ″

The cost of plywood, which is used extensively for subflooring, sheathing, and cabinet work, will depend on the thickness and grade of material and the location of the work and will vary from less than $100 to more than $250 msf. Some average prices are:

3/8″ standard grade	—	$100 msf
1/2″ ″ ″	—	125 ″
5/8″ ″ ″	—	150 ″
3/4″ ″ ″	—	180 ″

These approximate costs are for interior grade plywood; exterior grade plywood will be about 25% higher.

ROUGH CARPENTRY

Rough carpentry includes:
1. Framing

a.	Posts	f.	Sills	k.	Ceiling joists
b.	Girders	g.	Studs	l.	Trusses
c.	Floor joists	h.	Plates	m.	Rafters
d.	Bridging	i.	Caps	n.	Wind or collar beams
e.	Headers	j.	Ribbons	o.	Ridge piece, etc.

2. Sheathing and subflooring
3. Furring and grounds
4. Door and window frames (unless prehung items are used)
5. Insulation
6. Bracing

Framing Types

Figures 7-1 through 7-5 are illustrations of the two different systems of house framing (platform frame construction and balloon frame construction). Also identified are some of the items of carpentry that go into the makeup of a dwelling constructed principally of wood.

Taking Materials From Plans—Rough Carpentry

Sills. Where there are breaks in the foundation walls sills are used to bridge these gaps and to form a support for the sill plates. Sills can be 4″ × 6″, 6″ × 6″, 6″ × 8″, 6″ × 10″, or even larger. Five per cent should be allowed for end waste.

Sill Plates. Sill plates are placed on top of the foundation walls to support the joists and can be 2″ × 6″, 2″ × 8″, 2″ × 10″, 2″ × 12″, or larger.

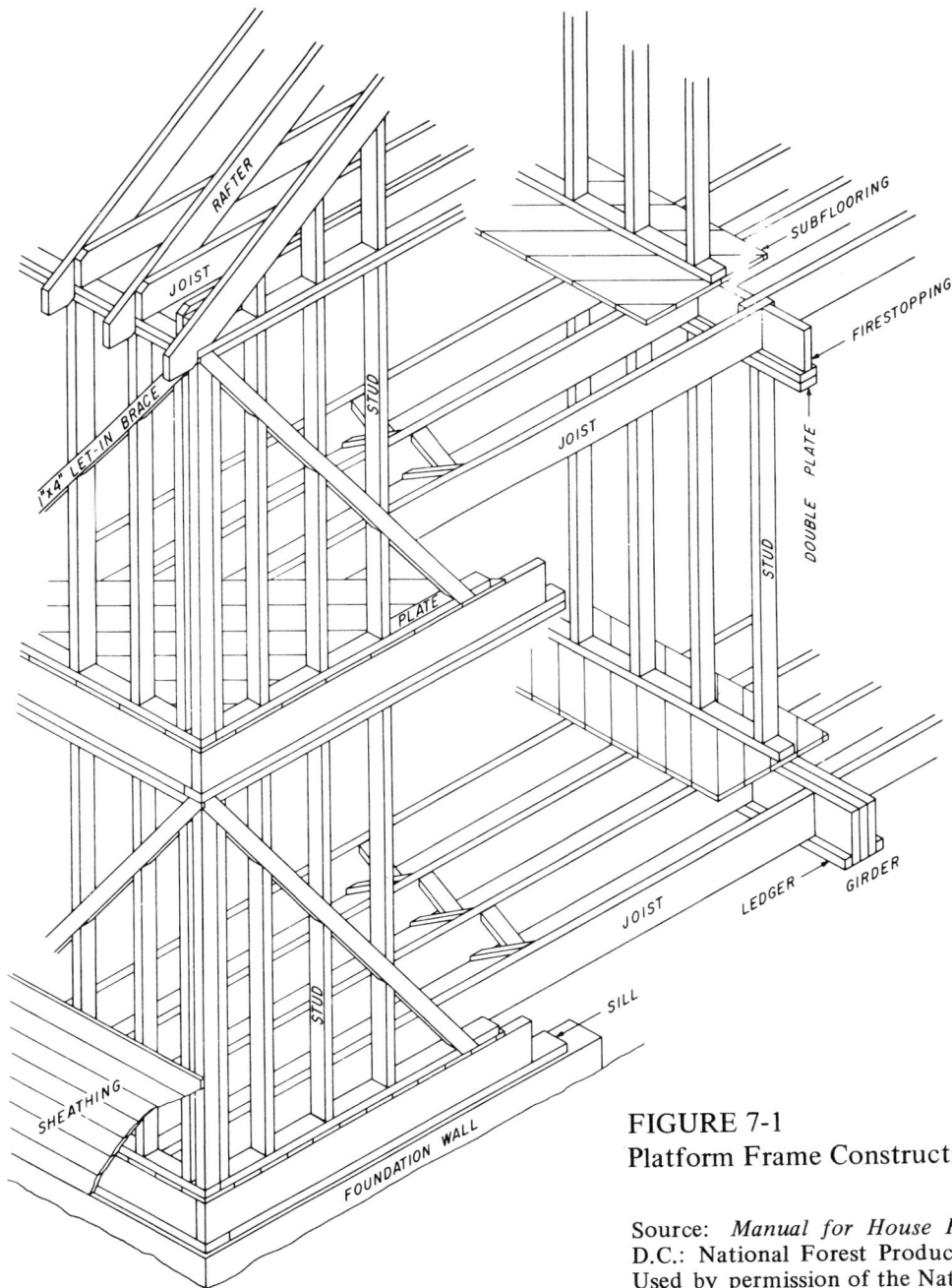

FIGURE 7-1
Platform Frame Construction

Source: *Manual for House Framing No. 1* (Washington, D.C.: National Forest Products Association, 1961), p. 13. Used by permission of the National Forest Products Association.

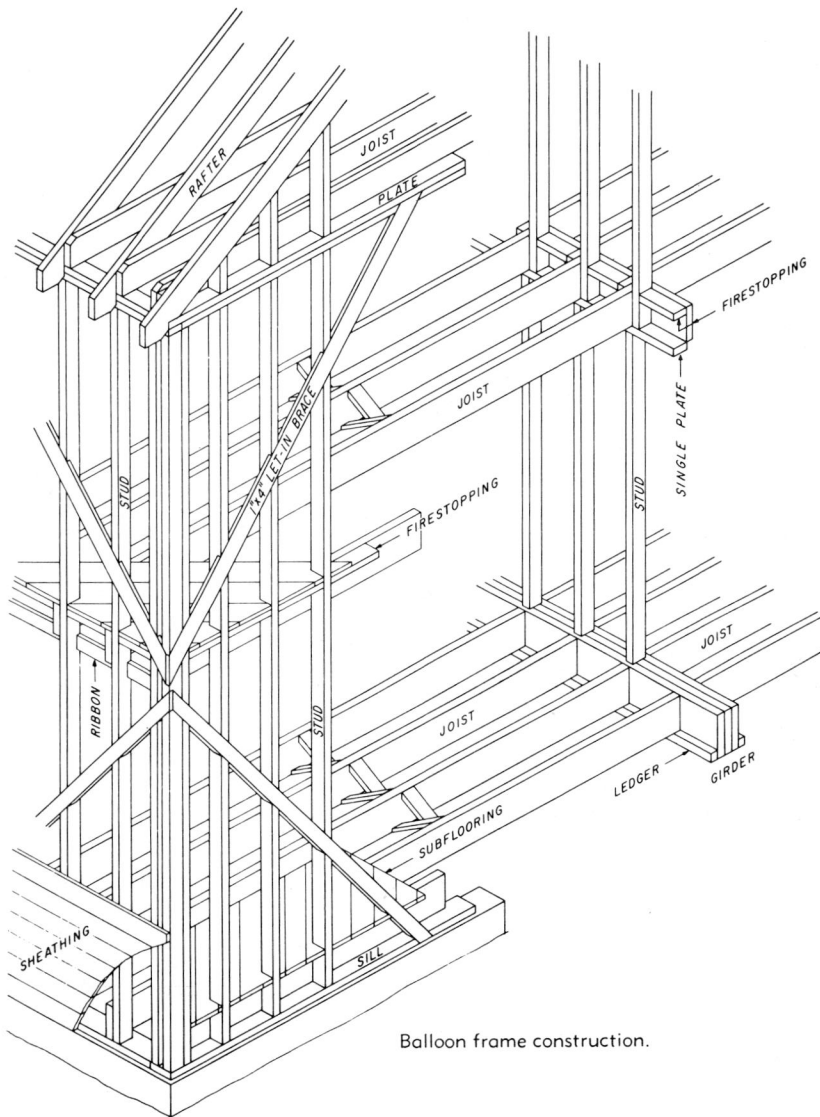

Balloon frame construction.

FIGURE 7-2
Balloon Frame Construction

Source: Manual for House Framing No. 1 (Washington, D.C.: National Forest Products Association, 1961), p. 14. Used by permission of the National Forest Products Association.

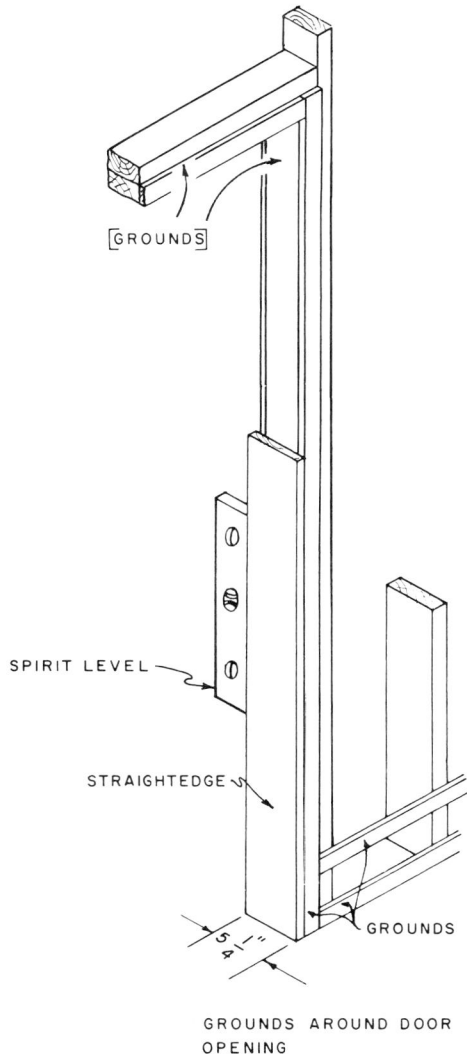

FIGURE 7-3
Illustration Showing Grounds Around
Door Opening

Source: Raymond P. Jones, *Framing, Sheathing and Insulation* (Albany, N.Y.: Delmar Publishers, 1964), p. 192.
Used by permission of Delmar Publishers.

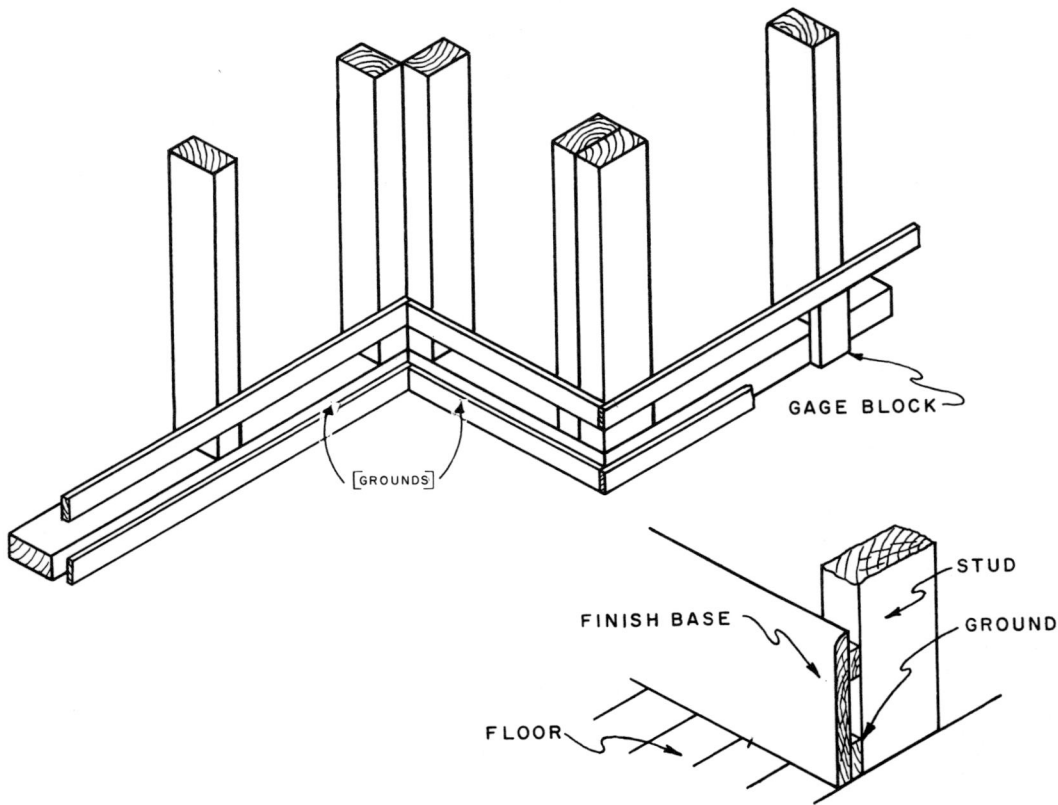

FIGURE 7-4
Illustration Showing Grounds Around Baseboard

Source: Raymond P. Jones, *Framing, Sheathing and Insulation* (Albany, N.Y.: Delmar Publishers, 1964), p. 191. Used by permission of Delmar Publishers.

Plates and Caps. Plates and caps are the same width as the studs and of two-inch material. There is one plate under the studs and there are two caps across the top.

Joists. The joists bear on the sill plate on the wall end and at the interior end are carried on the girder by direct bearing or by ledgers attached to the girder. These girders can be of one piece of wood but are usually made by fastening several pieces of 2″ × 8″, 2″ × 10″, 2″ × 12″ or 2″ × 14″ together.

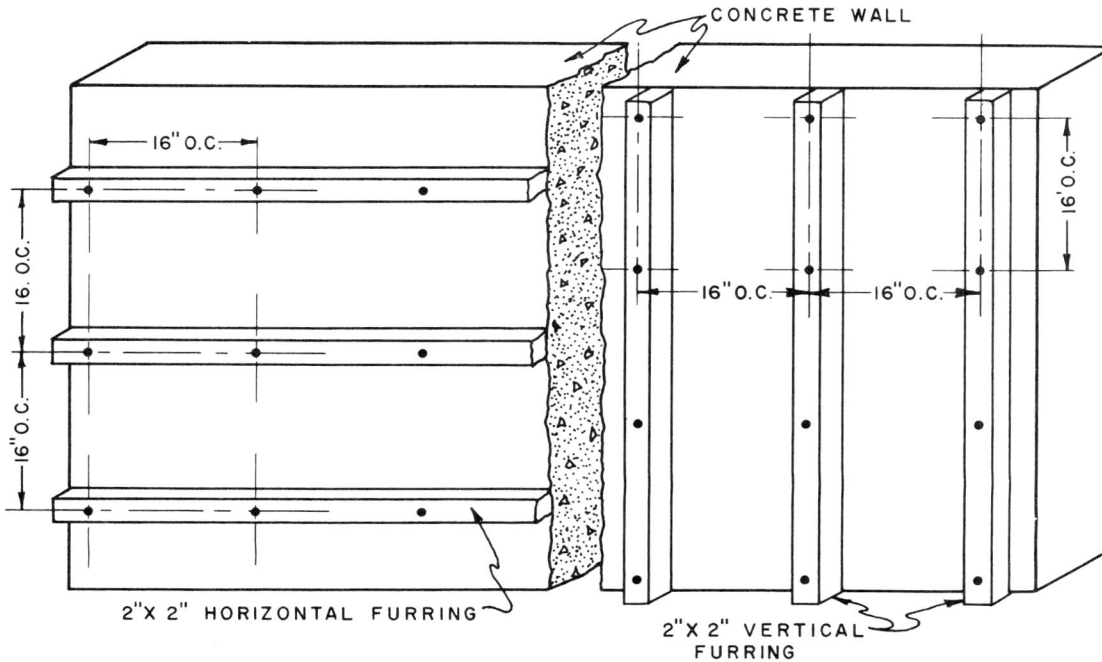

FIGURE 7-5
Illustration Showing Horizontal and Vertical Wall Furring

Source: Raymond P. Jones,*Framing, Sheathing and Insulation* (Albany, N.Y.: Delmar Publishers, 1964), p. 190. Used by permission of Delmar Publishers.

The floor joists are ordinarily doubled under a wall partition which runs parallel to the joists and at headers around stair wells and other openings in the floor.

Studs. Wall and partition studs are usually $2'' \times 4''$ and spaced on $16''$ centers though both size and spacing may vary.

Counting Studs and Plates. There is a convenient rule for finding the number of wall and partition studs and also plates and caps when the spacing is $16''$ on center:
1. Take 3/4 of the total lengths of all walls and partitions in feet.
2. Add one stud for each length of wall.

3. Add one stud where wall is intersected by a partition.
4. Add one stud for each length of partition.
5. Add one stud for each intersection of partitions.
6. Add two studs for each wall opening and two for each partition opening.
7. Add for plates and caps three times the total length of all walls and partitions divided by the length of the studs.
8. Allow 5% for waste.

Counting Joists. To find the number of joists when the spacing is 16″ on center:
1. Take 3/4 of the floor length perpendicular to the joists and add one joist.
2. Add one joist for each partition parallel to joist and one for each header.
3. Add 5% for waste.

Counting Rafters. To find the number of rafters when the spacing is 16″ on center:
1. Take 3/4 of the length perpendicular to the rafters and add one rafter.
2. Allow for the extra length due to the slope of the roof and for any overhang.
3. Allow 5% for waste.

Bridging. Take off at least one set of cross bridging between each two joists. Space lines of bridging in the range of 5 to 7 feet on centers. For joists on 16″ centers bridging can be figured as 1″ × 4″ × 18″. Allow 5% for waste.

Ribbons. Ribbons are usually 1″ × 4″ with one line around the walls required for each story with 5% allowed for waste.

Sheathing. Sheathing for walls and subflooring may be laid across the studs and joists or diagonally. Roofing is usually laid across the rafters. The material may be boards or plywood. One convenient method, which is on the safe side as to quantity for the walls, is to ignore all openings in taking off this material and figure that the extra quantity will adequately compensate for the waste. For subfloor and roof allow 15% for waste, if laid crosswise, and 20% if laid diagonally. If plywood is used cut wastage by 1/3.

Furring and Grounds. Furring and grounds are, except in rare instances, 1″ × 2″ or 1″ × 3″ material. It is usually sufficient to allow 2 linear feet of grounds for each foot of baseboard and for each foot of chair rail and one linear foot ·for each foot of picture mold and door and window casings. Furring must be measured from the plans. Allow 10% wastage for grounds and furring.

Door and Window Frames. Door and window frames can be set as the framing is being worked and would then be considered as rough carpentry. When door and window frames are used that come complete with prehung doors and windows, the installation will generally be done with the finish work; otherwise, the doors and windows would be in the way of the free movement needed during the rough work and would likely suffer damage during this period. For the discussion of doors and windows, see Chapter 10 (p. 209) on glazing.

Insulation. Insulation can be included in either rough or finish work. When the insulation is in batts to be placed between the studs or is in sheets to be placed on the outside of studs instead of sheathing or on the inside of studs instead of lathing, then the work is part of the rough carpentry. When the insulation is used as interior finish, the work is included as finish carpentry. In computing quantities at least 5% should be allowed for waste.

Labor for Rough Carpentry. Table 7-1 shows the approximate labor hours required for certain items of rough carpentry.

Equipment. The equipment for this type of work consists of electric hand and table saws, planers, etc., and a power outlet. An allowance of from $100 to $300 for an ordinary size structure in the $50,000 to $75,000 class should be sufficient. For scaffolding allow $3.00 to $4.00 per 100 sq ft of wall surface.

An allowance for nails of at least 20 lb per 1,000 fbm should be made. Nails will cost an average of $0.20 per pound.

Lumber Pricing Units. All lumber is priced per 1,000 fbm except for molding and certain articles of trim which are priced per foot.

Computing fbm. In computing the fbm of lumber for any material of a stated thickness go to the next highest inch because the finished lumber must be planed from rough lumber purchased in even inches.

TABLE 7-1
Approximate Man-Hours Required for Rough Carpentry

Item	Units	Man-Hours
Posts	1,000 fbm	15 - 25
Girders made from joists spiked together	1,000 fbm	25 - 35
Joists - 2 × 6; 2 × 8; 2 × 10	1,000 fbm	20 - 30
Joists - 2 × 12	1,000 fbm	14 - 24
Bridging	1,000 fbm	50 - 75
Sills	1,000 fbm	18 - 30
Studs - 2 × 4; 2 × 6	1,000 fbm	20 - 35
Plates, caps	1,000 fbm	18 - 38
Ribbons, 1 × 4	1,000 fbm	30 - 45
Ceiling joists - 2 × 4; 2 × 6; 2 × 8	1,000 fbm	24 - 36
Rafters	1,000 fbm	22 - 33
Small roof trusses	1,000 fbm	36 - 48
Subflooring* at 90°	1,000 fbm	15 - 22
Subflooring* diagonal	1,000 fbm	18 - 25
Sheathing* at 90°	1,000 fbm	15 - 25
Sheathing* diagonal	1,000 fbm	20 - 30
Wallboard	1,000 fbm	10 - 20
Furring on wood	100 lin ft	1 - 2
Furring on masonry	100 lin ft	2 - 3
Grounds	100 lin ft	3 - 5
Insulating batts between studs	100 lin ft	1 - 5
Insulating batts between joists	100 lin ft	1 - 4
Building paper	100 sq ft	0.5 - 1
Roofing	1,000 fbm	18 - 24
Placing outside door frames	Each	2 - 3
Placing inside door frames	Each	1.5 - 3
Placing single window frames	Each	1 - 2
Placing double window frames	Each	1.5 - 2.5

Labor cost per man may be computed by assuming a gang composed of:

4 Carpenters	@	$7.85 per hour	= $31.40
4 Helpers	@	6.15 per hour	= 24.60
1 Foreman	@	8.30 per hour	= 8.30 = $64.30

Cost per man-hour $64.30 / 9 = 7.14

*Man-hours will be 25% less when plywood is used.

Since lumber transactions are carried on in fbm (foot board measure) it is important to know what constitutes a board foot. A board foot is a square foot of lumber one inch thick. One of the best ways to compute fbm is to set an equation with the numerator as the section of the piece in inches and the denominator as 12 all multiplied by the length of the board in feet. This will automatically give the answer in square feet of material so many inches thick; otherwise care must always be exercised to keep one of the dimensions of the section in inches while converting the other to feet while using the length in feet. Two examples follow:

2 pieces of 3/4″ × 6″ × 10′ - 0″ = 2(1 × 6) /12 × 10 = 10 fbm
3 pieces of 1 1/4″ × 8″ × 12′ - 0″ = 3(2 × 8) / 12 × 12 = 48 fbm

Standard Lumber Lengths. In taking lumber from the plans for an estimate it should be borne in mind that lumber comes from the mill in even feet increments such as 2, 4, 6, 8, 10, 12, 14, 16, 18, 20, etc., unless otherwise specifically ordered. In order to get a joist 13′ - 6″ long it would be necessary to order a piece of 14′ in length unless a special length were ordered at an extra cost per fbm.

FINISH CARPENTRY

Finish carpentry includes:
Interior Finish
1. Floors
   a. Flooring
   b. Building paper
   c. Deadening felt
   d. Sanding
2. Ceilings and walls (for lathing and plastering and drywall construction, see chapter 11, p. 215).
3. Windows
4. Doors (for doors and windows, see chapter 10, p. 209).
5. Trim
   a. Baseboard
   b. Shoemold
   c. Chair rail
   d. Picture mold
   e. Crown mold
   f. Wainscot

g. Paneling

h. Beam casings

6. Stairs

    a. Stringers

    b. Treads

    c. Risers

    d. Railing

7. Millwork

    a. Cabinets

    b. Closets

    c. Bookcases

    d. Columns

    e. Mantels

    f. Shelving, etc.

8. Interior hardware

Exterior Finish

1. Porches or stoops

    a. Framing

      1. Floor

      2. Roof

    b. Sheathing

      1. Floor

      2. Roof

    c. Finishing

      1. Flooring: wood, concrete, tile

      2. Roofing

    d. Steps

      1. Wood

      2. Concrete

2. Exterior walls

    a. Building paper

    b. Siding

3. Doors

4. Windows

5. Trim

    a. Cornice

    b. Soffits

    c. Corner boards

    d. Water tables

6. Roofing
  a. Covering
    1. Wood or composition shingles
    2. Tile
    3. Tin
  b. Flashing (see chapter 9, p. 203)
    1. At valleys
    2. At chimneys
    3. Gutters and downspouts
7. Exterior hardware and vents

Taking Materials From Plans—Finish Carpentry

Species of Wood. Some of the several species of wood that find their way into the finished floor surfaces are: pine, fir, maple, oak, beech, and birch. The softwoods, which include pine and fir, are cheaper than the hardwoods and require less labor to install (as shown in Table 7-2.)

Finish Flooring. In computing the fbm of lumber necessary to provide finish flooring considerable wastage must be allowed due to the tongue and groove effect and end cuts. Table 7-2 will take into account these losses and give some idea of the fbm required per 100 square feet of floor area.

TABLE 7-2
Finish Flooring Requirements
per 100 Sq Ft of Surface

Nominal size (in.)	Type	Fbm per 100 sq ft	Labor hours per 100 sq ft	Nails per 100 sq ft (in lb)
1 × 2	Softwood	135	3.0 - 4.3	10
1 × 3	Softwood	125	2.7 - 4.0	8
1 × 4	Softwood	122	2.4 - 3.7	8
1 × 5	Softwood	118	2.0 - 3.3	6
1 × 6	Softwood	115	1.6 - 2.7	6
1 × 2 1/4	Hardwood	150	4.0 - 5.3	10
1 × 2 1/2	Hardwood	145	3.7 - 5.0	8
1 × 2 3/4	Hardwood	140	3.3 - 4.6	8
1 × 3	Hardwood	135	2.9 - 4.3	8
1 × 4	Hardwood	125	2.7 - 4.0	6

Hardwood flooring will vary in cost from about $275 to $600 per fbm depending upon species, size of board, and grade. Softwood flooring will cost 10 to 15% less than hardwood.

Bevel Siding. In estimating the required amount of bevel siding an allowance has to be made for the wastage due to laps and end cuts. Table 7-3 shows approximate ranges for this allowance.

Siding will cost somewhere in the range of $170 to $350 per mfbm depending on the kind of wood and grade specified.

TABLE 7-3
Siding Required per 100 Sq Ft of Area

Width (in inches)	Fbm per 100 sq ft	Nails per 100 sq ft (in lb)
6	125 to 135	3 - 4
5	145 to 155	3 - 4
4	155 to 165	4 - 5

Roofing. Roofing is priced by the square, which means a 10 foot by 10 foot space or 100 square feet (see chapter 9, p. 203, for discussion of wood and asphalt shingles).

Molding and Trim. Molding and trim are priced by the linear foot and require 2 to 4 pounds of nails per 100 linear feet. About 10% should be allowed for waste. (One-inch thick molding will cost approximately $0.05 per inch of height per foot.)

Kitchen Cabinets. Kitchen cabinets are priced in most cases by the linear or the cubic foot. The price will depend on the type of cabinets designated.

Doors and Windows. The price of doors and windows will vary due to size, type, and material. When prehung by the mill they will cost more, but there will be a considerable saving in the labor on the job.

The price of kitchen cabinets, doors, windows, and trim will vary a great deal from one locality to another and the local price for each item must be ascertained.

Labor for Finish Carpentry. Tables 7-4 and 7-5 show the approximate labor hours required for certain items of interior and exterior finish carpentry.

TABLE 7-4
Approximate Man-Hours Required for
Interior Finish Carpentry

Item	Units	Man-Hours
Deadening felt on subfloor	100 sq ft	0.4 - 1.1
Building paper on subfloor	100 sq ft	0.3 - 1.0
Sanding floors	100 sq ft	1 - 3
Paneling	100 sq ft	4 - 14
Plaster board	100 sq ft	1 - 3
Wainscot	100 sq ft	3 - 5
Baseboard	100 lin ft	5 - 7
Chair rail	100 lin ft	4 - 7
Molding	100 lin ft	2 - 6
Doors (single with hardware)	each	4 - 6
Doors (double with hardware)	each	5 - 9
Doors (prehung in frame)	each	2 - 4
Stairs	per ft height	3 - 8
Mantels	each	4 - 10
Millwork such as: cabinets, broom closets, ironing boards, bookcases, etc.	each	2 - 8
Floors (see Table 7-2 for flooring labor)		

Note: See chapter 11 for drywall construction, lathing, and plastering.

TABLE 7-5
Approximate Man-Hours Required for Exterior
Finish Carpentry

Item	Units	Man-Hours
Siding	100 sq ft	2 - 8
Shingles on walls	100 sq ft	3 - 7
Building paper on walls	100 sq ft	1 - 2
Shingles on roof	100 sq ft	2 - 5
Roofing felt	100 sq ft	1 - 2
Fascias	100 sq ft	8 - 20
Cornice	100 lin ft	3 - 21
Blocking	1,000 fbm	50 - 80
Corner pieces	100 lin ft	2 - 4
Water tables	100 lin ft	3 - 5
Shutters	pair	1 - 3
Porches:		
Ceiling	100 sq ft	3 - 6
Columns	each	1 - 3
Railing (wood)	100 lin ft	35 - 75
Steps	each set	7 - 15
Windows (single with hardware)	each	4 - 7
Windows (single—weather stripping)	each	1 - 2
Windows (double with hardware)	each set	6 - 9
Doors (single with hardware)	each	5 - 8
Doors (single—weather stripping)	each	2 - 3
Doors (double with hardware)	each set	6 - 12
Doors and windows (single prehung in frames)	each	2 - 5
Screen doors and windows	each	1 - 2

EXAMPLE 7-1

For Example 7-1 we will use by permission a set of house plans (Figure 7-6) prepared by Robert B. Lyons, an architect in Raleigh, N.C. These plans were to be used in a low-cost housing project so are not on the elaborate side.

The purpose of this example is to illustrate the estimating of rough and finish carpentry, but in order to obtain a price by which we can arrive at a square foot figure the cost of the clearing, grubbing, excavating, foundations, chimney, siding, brick veneer, finishes for walls, floor and ceiling finishes, plumbing, heating, air conditioning, electrical, and painting will be added.

The remaining chapters in this text are to be studied as required by each material not yet covered as it appears in Example 7-1. This will be the last example estimate shown.

[The house plans (Figure 7-6) appear on pp. 158-161. Quantity Takeoff, Summary & Unit Cost, Direct Cost, and Overhead & Profit forms of Example 7-1 appear on pp. 162-188. Text of Homework Example 7-1 is on p. 189.]

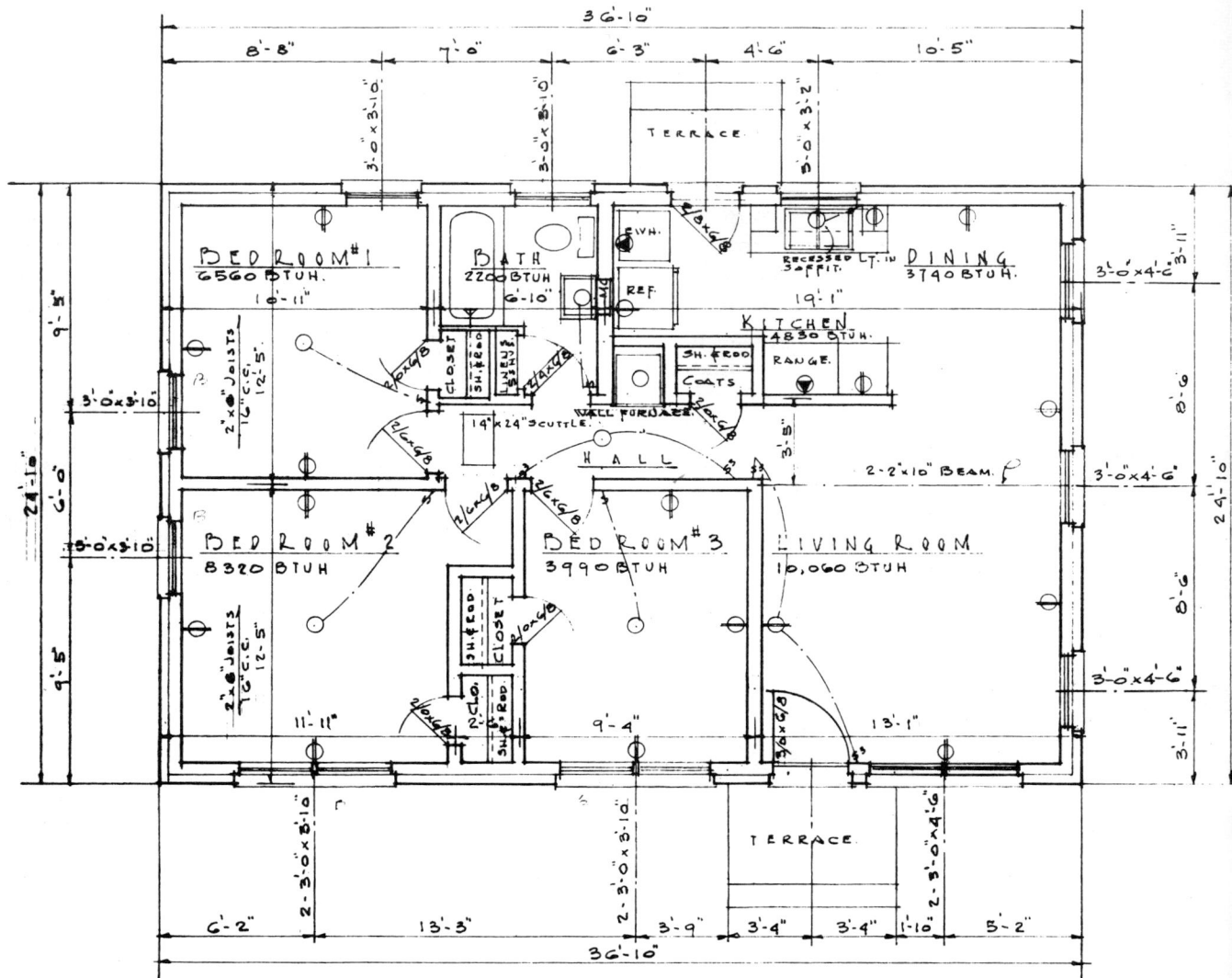

FIRST FLOOR PLAN

SCALE - 1/4"=1'-0"

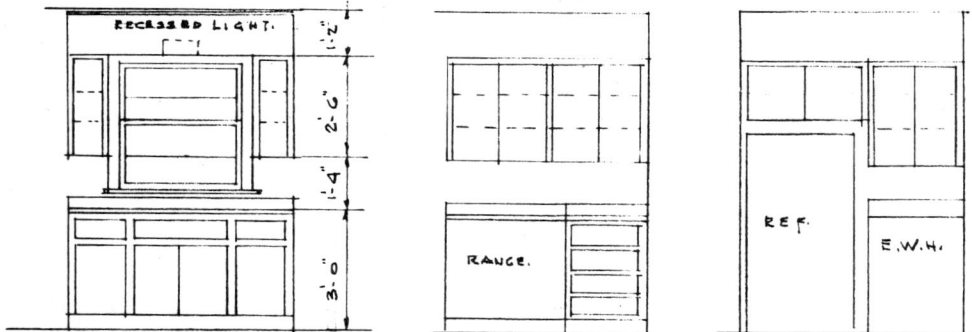

KITCHEN INCLUDES
BASE CABINETS - 7.5 LIN.FT.
WALL CABINETS - 36.5 LIN.F
DRAWERS - 8.0 LIN.FT.

FIGURE 7-6
Blueprints for a House

KITCHEN CABINET DETAILS

SCALE 3/8"=1'-0"

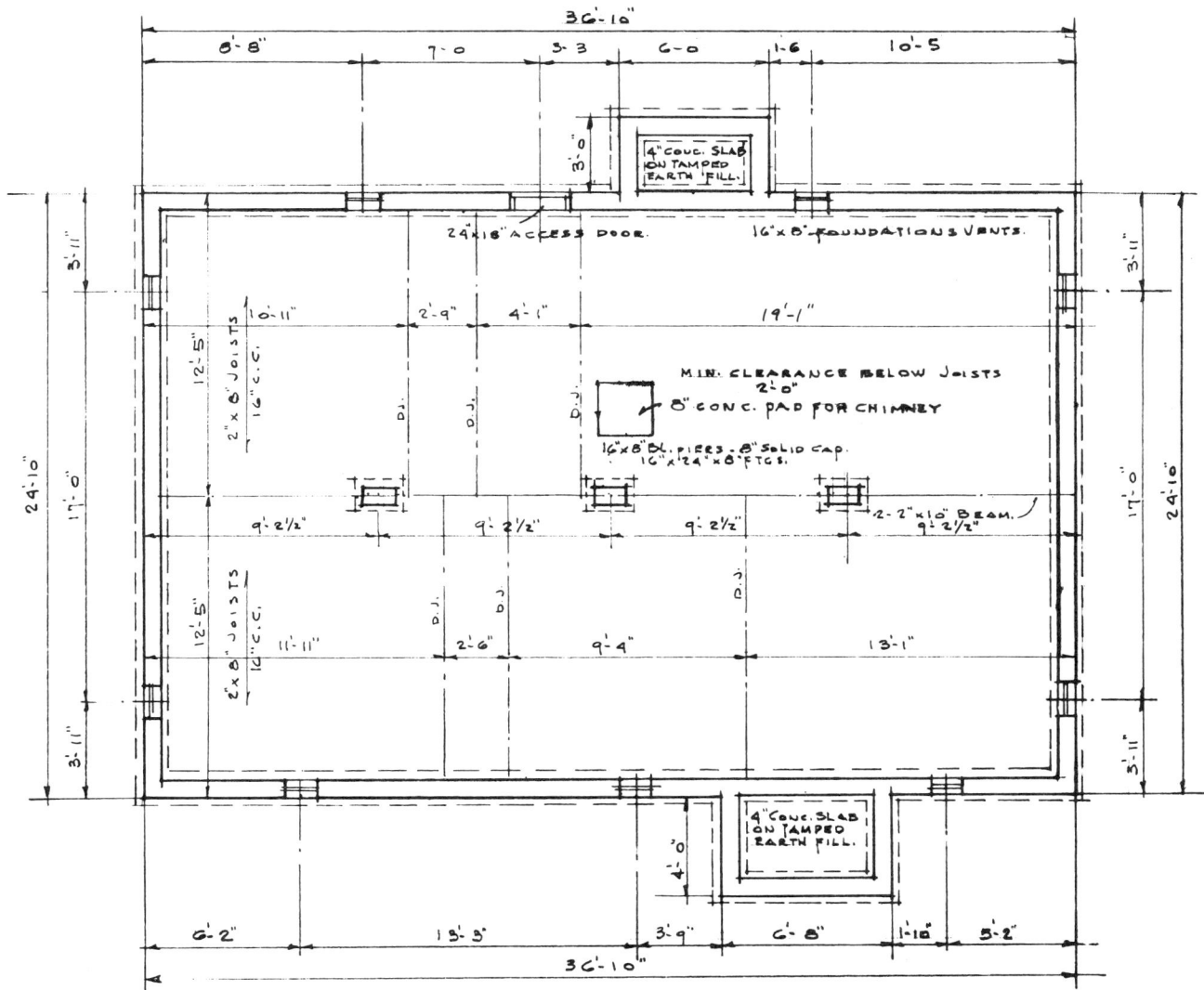

FOUNDATION PLAN
SCALE-1/4"=1'-0"

FIGURE 7-6 (continued)

CRAFT HOMES INC.
RALEIGH N. CAROLINA

ROBERT B. LYONS
ARCHITECT
RALEIGH, N. CAROLINA.

159

LEFT SIDE ELEVATION
SCALE: 1/4"=1'-0"

RIGHT SIDE ELEVATION
SCALE: 1/4"=1'-0"

REAR ELEVATION
SCALE: 1/4"=1'-0"

FRONT ELEVATION
SCALE: 1/4"=1'-0"

160

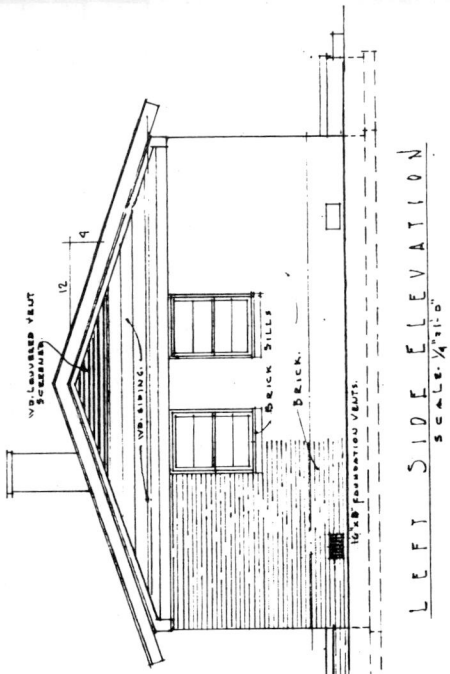

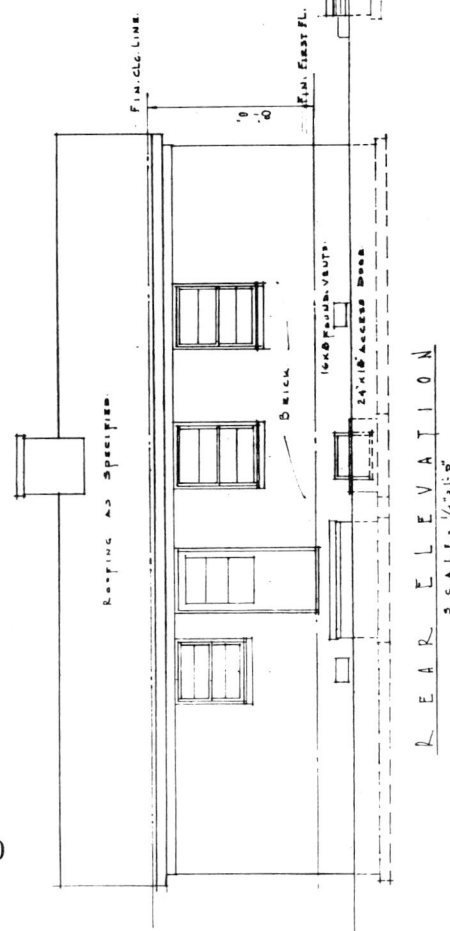

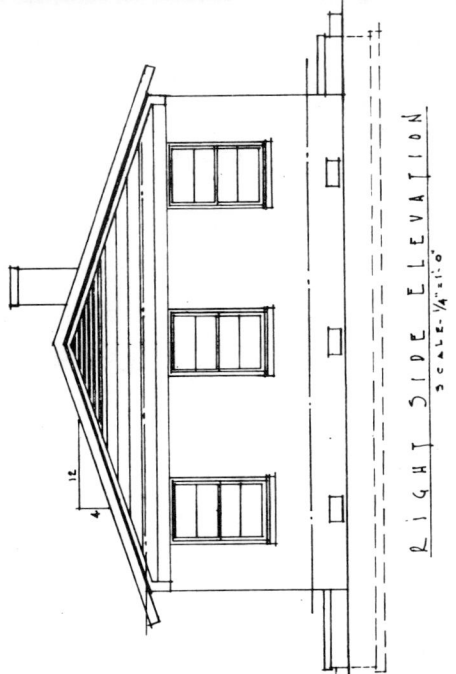

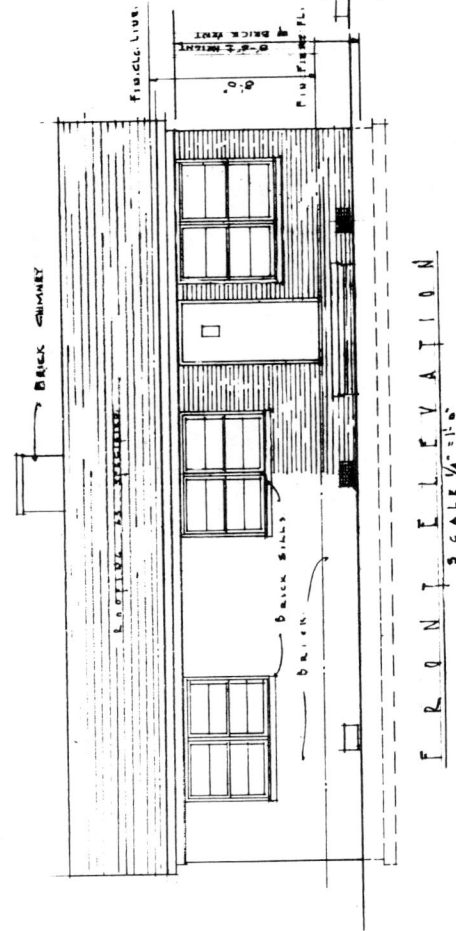

FIGURE 7-6 (continued)

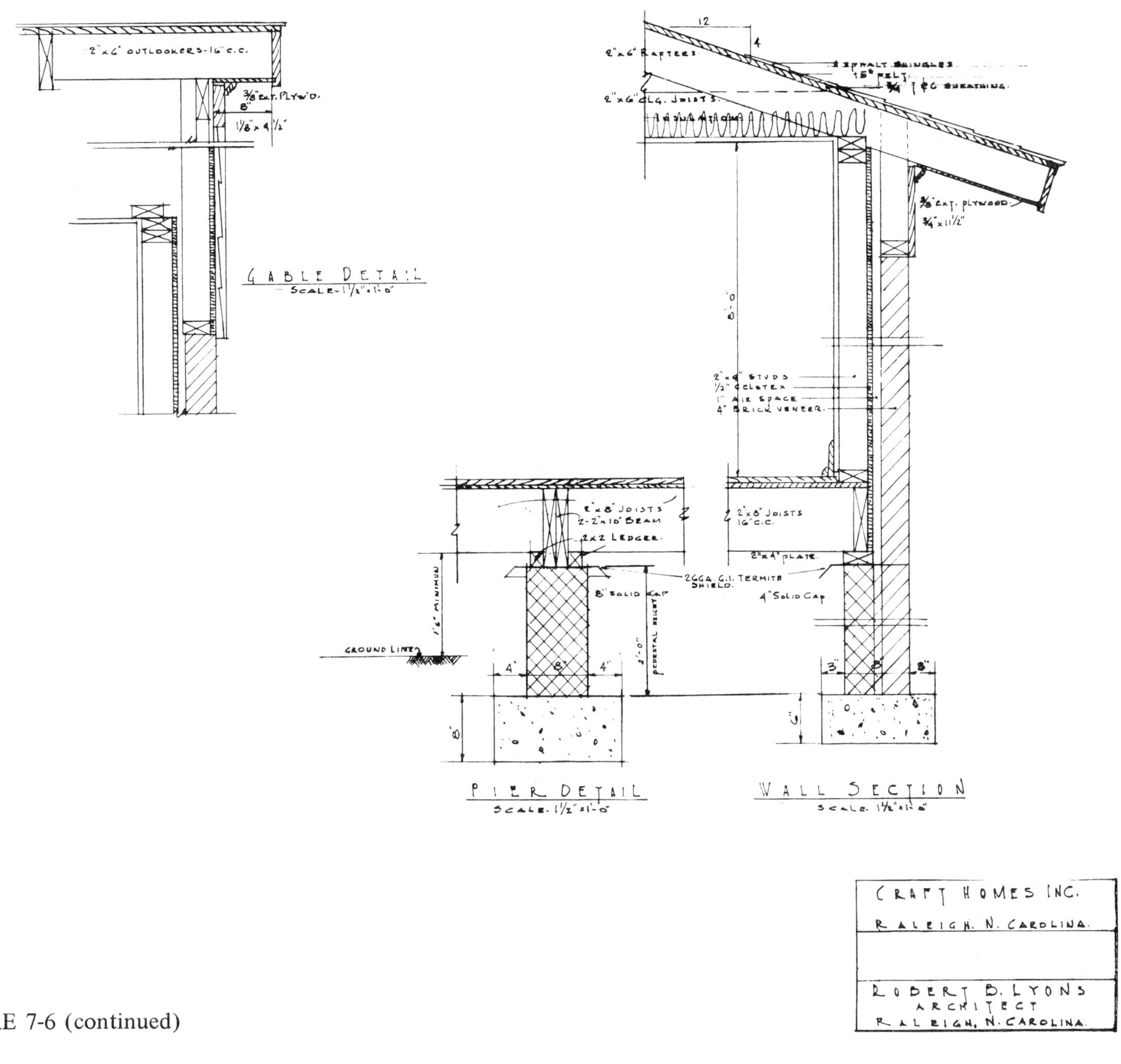

WALL SECTION
SCALE: 1/2"=1'-0"

PIER DETAIL
SCALE: 1½"=1'-0"

GABLE DETAIL
SCALE: 1½"=1'-0"

CRAFT HOMES INC.
RALEIGH, N. CAROLINA.

ROBERT B. LYONS
ARCHITECT
RALEIGH, N. CAROLINA.

FIGURE 7-6 (continued)

161

QUANTITY TAKEOFF—Example 7-1

Estimate by _____   Date _____   Ckd. by _____   Date _____

Item	Identity	Location	Quantity Computations	Total	Units
300	Clearing and grubbing			$300	Lump Sum
301	Excavating	Foundations		$150	Lump Sum
400	Concrete	Wall foot	[(36.17 + 24.17)2 + (2 x 2.33 + 6) + (2 x 3.33 + 6.67)] x 0.5 x 1.17/27	3.2	cu yd
	"	Chimney foot	3 x 3 x 1./27	0.4	cu yd
	"	Piers	(1.33 x 2.0 x 0.67) x 3/27	0.2	cu yd
600	8" reg. conc. block	Foundation walls	Allow 2 courses of 8" block and one courses of 4" [(36.17 + 24.17)2 + (2 x 2.33 + 6) + (2 x 3.33 + 6.67)] ÷ (1.33 x 2 x 1.03)	2.3	100
602	4" solid conc. block	Foundation walls	[(36.17 + 24.17)2 + (2 x 2.33 + 6) + (2 x 3.33 + 6.67)] ÷ (1.33 x 1.03)	1.2	100
600	8" reg. conc. block	Piers	Allow 2 courses of 8" reg. conc. block topped with one row of 8" solid block	0.1	100
601	8" solid conc. block	Piers		0.1	100
610	Standard size face brick	Outside walls	[8.5(36.5 + 24.5)2 - (3 x 6.67)2 - (3 x 4.5)14]7 x 1.05	6.	1000
612	Standard size face brick	Chimney	18[2 x 2 - (0.67 x 0.67)]21.4 x 1.05	1.5	1000
611	Metal ties	Walls	[8.5(36.5+24.5)2- (3 x 6.67)2 - (3. x 4.5)14	818	Each
		ROUGH	CARPENTRY		
700	Beams	Floor	2(2" x 10"/12") 18' x 2 x 1.05	126	fbm
		Ceiling at living room	2(2" x 10"/12") 14' x 1.05	50	fbm
701	Equipment	Carpentry	Small tools, etc.	1	Lump sum

Estimate by _____ Date _____ Ckd. by _____ Date _____

Item	Identity	Location	Quantity Computations	Total	Units
702	Ledgers	Beams	2(2" x 2"/12") 36 x 1.05	26	fbm
703	Joists	Floor	2[(2" x 8"/12") 12'(36' x 3/4 + 1)]1.05	941	fbm
703	Extra joists	Under partitions	(2" x 8"/12") 12' x 8 x 1.05	135	fbm
704	Headers	At ext. walls	(2" x 8"/12") (36' + 24') 2 x 1.05	168	fbm
		doors & windows	(2" x 4"/12") (3' x 2 x 16) x 1.05	67	fbm
705	Bridging	At joists	[(1" x 4"/12") 1.5 (36 x 3/4] 2 x 1.05	29	fbm
706	Studs	Ext. walls	2(36' + 24') 3/4 + 4 + (2 x 16) = 126		
		Int. walls	(140 x 3/4) + 14 + (2 x 9) = 137		
			Total number of studs = 263		
			fbm of studs = 8(2" x 4"/12")		
			263 x 1.05 =	1,400	fbm
707	Plates and caps	At studs	[2(36' + 24') + 140']3x		
			(2" x 4"/12") x 1.05 =	546	fbm
		At foundation	2(36' + 24') (2" x 4"/12") 1.05 =	84	fbm
708	Ribbons	In ext. walls	(1" x 4"/12") 2(36' + 24')	40	fbm
709	Joists	Ceiling	2[(2" x 6"/12") 14'(36' x 3/4 + 1)]x		
			1.05 =	833	fbm
710	Rafters	Roof	2[(2" x 6"/12") 16'(36' x 3/4 + 1)]x		
			1.05 =	952	fbm
	Ridgepiece	Roof	(2" x 10"/12") 20' x 2 x 1.05 =	73	fbm
	Collar beams	Roof	(2" x 6"/12") 8' x 9' x 1.05 =	80	fbm
	Outlookers	Roof (ends)	4(2" x 6"/12") 2' x (15' x 3/4)1.05 =	93	fbm
711	Termite shield		2(36' + 24') + 3(3')	129	S.F.
712	Subfloor	Floor	(36' x 24') x 1.10 (1/2" plywood) = 993	1,000	S.F.
			use		
713	Sheathing	Roof	2(15' x 38') 1.15 (3/4" T. & G)	1,311	fbm
714	Insulation	Ceiling	(36' x 24') 4" thick)	9	100 sqft
715	Celotex	At brk. walls	9' (36' + 24') 2 x 1.10 (Neglect open-	13.3	100 sqft
	(Sheathing)	Undersiding	ings) (24' x 4') x 1.5 (for waste)		
		FINISH CARPENTRY – INTERIOR			
717	Hardwood	Floor	(36' x 24') – (6' x 18')	7.6	100 sqft.
718	Bldg. paper	Floor	(36' x 24')	9.0	100 sqft

Estimate by _____ Date _____ Ckd. by _____ Date _____

Item	Identity	Location	Quantity Computations	Total	Units
1100	Gypsum board (Dry wall)	Ceiling	(36' x 24') x 1.10 (For waste)	9.5	100 sqft
1101	Gypsum board (Dry wall)	Ext. Walls Int. Walls	8 (36' + 24')2 (No allowance 8(73)2 for openings)	21	100 sqft
1102	Ceramic tile	Bath floor	(4' x 8')	32	sq ft
1103	Ceramic tile	Bath walls	(8' x 4')3.5' + (2.5' + 2.5' + 5')5'	92	sq ft
1104	Vinyl tile	Kitchen floor	(8' x 11')	88	sq ft
1000	Windows		3' - 0" x 4' - 6" double hung wood	14	Each
1010	Doors	Front door	3/0 x 6/8 x 1 3/4 solid core Single light prehung	1	Each
1011	Doors	Rear door	2/8 x 6/8 x 1 3/4 solid core 3 lights prehung	1	Each
1015	Doors	Interior	2/8 x 6/8 x 1 3/8 hollow core	3	Each
1016	Doors	Interior	2/4 x 6/8 x 1 3/8 hollow core	1	Each
1017	Doors	Interior	2/0 x 6/8 x 1 3/8 hollow core	5	Each
720	Int. trim	Baseboard Shoe mold Crown mold		200 200 250	Lin ft Lin ft Lin ft
725	Kitchen-	Base	24" deep x 36" high	7.5	Lin ft
726	Cabinets	Wall	12" deep x 30" high	13.0	Lin ft
730	Closets	Clothes	3 shelves & rods each	4	Each
731	Closets	Linen	5 shelves	1	Each
735	Hardware	Thresholds	3' long bronze saddles	2	Each
		FINISH CARPENTRY - EXTERIOR			
410	Conc. terrace	Entrances	(6.67 x 6 + 6 x 4) x 0.33/27	1	cu yd
740	Bevel siding	At ends	(24' x 4') x 1.20 (Waste due to triangular shape)	1.15	100 sqft
741	Bldg. paper	At ends		115	sq ft

QUANTITY TAKEOFF—Example 7-1 (continued)

Estimate by _____ Date _____ Ckd. by _____ Date _____

Item	Identity	Location	Quantity Computations	Total	Units
745	3/8" ext. plywood	Soffit at roof	(1.33'(38' x 2) + 0.67'(32' x 2))1.20 =	.2	1000sqft
746	3/4"x11 1/2"Wd.	Fascia	(1" x 12")/12"(38' x 2) x 1.10	84	fbm
	1 1/8"x4 1/2" Wd.	Fascia	(2" x 12")/12"(32' x 2) x 1.10	141	fbm
	2" x 6" Wd.	Ends roof	(2" x 6")/12"(38' x 2) x 1.10	84	fbm
	2" x 6" Wd.	Ends roof	(2" x 6")/12"(32' x 2) x 1.10	70	fbm
747	2" Molding	At overhang	(38' x 2 + 16' x 4) x 1.10	150	Lin ft
750	Shingles–Asph.	Roof	15' x 2 x 38.33' x 1.10	12.7	100 sqft
751	15# roof. felt	Roof		12.7	100 sqft
900	Flashing	Chimney	16 oz copper	20	sq ft
901	Flashing	Ridge strip	26 ga galv. steel	38	Lin ft
902	Gutters	Front & Rear	5" aluminum	76	Lin ft
903	Downspouts	Each corner	5" aluminum	40	Lin ft
1110	Varnish	Hardwood Floors	(See item 717)	7600	sq ft
1111	Sanding	do.		7600	sq ft
1112	Painting	Walks & ceiling	(See items 1100 & 1101) Primer & 2 coats of latex	3050	sq ft
1113	Painting	Kitchen	(11 x 8) + (11 + 11 + 8)8 1 Primer & 2 coats enamel	330	sq ft
1114	Painting	Bath	(8 x 7) + (8 + 8 + 7 + 7)L 1 Primer & 2 coats enamel	180	sq ft
1115	Painting	Windows	1(3 x 3) + 8(3 x 4) + 5(3 x 4.5) 1 Primer & 2 coats enamel	175	sq ft
1116	Painting	Doors	1(4 x 7) + 4(3.5 x 7) + 1(3 x 7) + 5(3 x 7) 1 Primer & 2 coats enamel	250	sq ft
1117	Painting	Baseboard	(See item 720)	200	Lin ft

Estimate by _____ Date _____ Ckd. by _____ Date _____

Item	Identity	Location	Quantity Computations	Total	Units
	(Int. trim)	Shoemold	do.	200	Lin ft
		Crownmold	do.	250	Lin ft
			1 Primer & 2 coats enamel		
1118	Painting woodwork	Inside of closets and inside of kitchen cab.	$2[(3 + 2)2 \times 8] + 1[(3 + 1 + 1)8 + 4(3 \times 1)2] + 2[(4 + 2)2 \times 8]$	420	sq ft
				400	sq ft
			1 Primer & 2 coats latex		
1119	Painting	Wood Siding	(See item 740) 1 Primer & 2 coats oil base	115	sq ft
1120	Painting (ext. trim)	Plywood Soffit	See item 745	200	sq ft
		Fascia	See item 746 – 84 fbm 1" thick =	84	sq ft
		Fascia	See item 746 –141 fbm 2" thick =	71	sq ft
		Ends rafters	See item 746 – 84 fbm 2" thick =	42	sq ft
		Ends roof	See item 746 – 70 fbm 2" thick =	35	sq ft
		2" Molding	See item 747 –150 fbm 2" thick =	300	sq ft
760	Painting equipment			1	L. S.
765	Vents	Foundations	6" x 8" Cast iron	10	Each
1500	Heating and air conditioning			1	L. S.
1510	5' P.E. Tub with shower and curtain			1	Each
1511	Lav. wall hung 19" x 17" P.E.			1	Each
1512	W.C. vitreous china stand. type			1	Each
1513	Medicine cab.			1	Each
1514	Water heater – 80-gal elec. glass lined			1	Each

Estimate by ———————— Date ———————— Ckd. by ———————— Date ————

Item	Identity	Location	Quantity Computations	Total	Units
1515	Kitchen sink - single bowl P.E. one drain board			1	Each
1516	1" Copper service from main			75	Lin ft
1517	4" C.I. sewer pipe			75	Lin ft
1520	Electrical			1	L. S.

SUMMARIES & UNIT COSTS—Example 7-1

Estimate by _____ Date _____ Ckd. by _____ Date _____

Item No.	Identity & Cost Source	Computation of Unit Costs	Total	Units
300	Clearing and grubbing		$ 300.	L. S.
301	Excavating foundations		150.	L. S.
400	Conc. footings		3.8	cu yd
		Cost per cu yd.: Material cost = $20.00		
	Ex. 4-4	Placing conc. Labor cost = $ 4.00		
600	8" Regular conc. block (Hvy agg)	Cost per 100:	2.4	100
		Block cost (Table 6-7) = $31.50		
	3/8" Joints	Cement (Ex. 6-1) = $ 4.32		
	1-3 Mortar mix	Lime do. = $ .68 Sand do. = $ 1.20 Material cost $37.70		
		Mason & helper = 8.40 + 6.20 = $14.60 Labor 5 hrs/100 @ $14.60 x 1.10 = $80.30		
	Table 6-4			
601	8" solid conc. block (hvy agg)		.1	100
		Cost per 100: Block cost /100 (Table 6-7) = $58.50 Cement (Ex. 6-1) = $ 4.32 Lime do. = $ .68 Sand do. = $ 1.20 Material cost $64.70		
		Labor (same as above) = $80.30		

Estimate by _____ Date _____ Ckd. by _____ Date _____

Item No.	Identity & Cost Source	Computation of Unit Costs	Total	Units
602	4" Solid conc. block (hvy agg)	Cost per 100: Block cost/100 (Table 6-7) = \$ 30.60 Cement (Ex. 6-1) = \$ 4.32 Lime do. = \$ .68 Sand do. = \$ 1.20 Material cost \$ 36.80	.1	100
	Table 6-4	Labor 3.8 hrs/100 @ 14.60 x 1.10 = \$ 61.03		
610	stand. size face brick 1/4" Joints 1-1-5 Mortar mix Table 6-6 Table 6-1 do. do.	Cost per 1000: (See method of Ex. 6-1) Brick = \$ 70.00 Cement 0.33 (5.4 x \$1.60/sack) = \$ 2.85 Lime 0.33 (211/2000 x \$50./ton) = \$ 1.74 Sand 0.33 (1 x \$4.00/cu yd) = \$ 1.32 Material cost \$ 75.91	.6	1000
	Table 6-4	Mason & helper = 8.40 + 6.20 = \$ 14.60 Labor = 13 hrs/1000 @ 14.60 = \$189.80		
	Table 6-9	Mortar boxes, mixer etc. = \$ 3.00 Small tools & misc. equip. = \$ 1.00 Scaffold = \$ 6.00 Misc. = \$ 3.00 Equipment \$ 13.00		
611	Metal ties	818 Cost = \$0.02 each	818	Each

Estimate by _____ Date _____ Ckd. by _____ _____ Date _____

Item No.	Identity & Cost Source	Computation of Unit Costs	Total	Units
612	Face brick in chimney		1.5	1000
		Cost per 100:		
	1/4" Joints 1-1-5 Mortar mix	See item #610 for Material cost = $ 75.91		
	Table 6-2	Mason & helper = $8.40 + 6.20 = $ 14.60 Labor = 26 hrs/1000 @ 14.60 = $379.60		
	Table 6-9	Mortar boxes, mixer etc. = $ 3.00 Small tools & misc. equip. = $ 1.00 Scaffold = $ 12.00 Misc. = $ 3.00 Equipment $ 19.00		
	Rough Carpentry			
700	Girders made of two 2 x 10	126 + 50 = 176 fbm cost per mfbm:	.18	mfbm
	See Mat'ls - 7 * See Equip. - 7	Lumber = $160.00 Nails - 20 lb/mfbm @ 0.20 = $ 4.00 Material = $164.00		
	Table 7-1	Labor - 25 hr @ $7.14 = $178.50		
701	Equipment for carpentry	Saws, planers, power outlet, etc. Scaffolding @ $4.00/100 sq ft = $4. x 10	$75. $40.	L. S. L. S.
702	Ledgers	26 fbm = Cost per mfbm:	.03	mfbm
	See Mat'ls - 7 See Equip. - 7	Lumber = $150.00 Nails - 30 lb/mfbm @ 20 = $ 6.00 Material = $156.00		
	Table 7-1	Labor - 40 hr @ $7.14 $285.60		
		* "7" Refers to Chapter 7.		

SUMMARIES & UNIT COSTS—Example 7-1 (continued)

Estimate by _____ Date _____ Ckd. by _____ Date _____

Item No.	Identity & Cost Source	Computation of Unit Costs	Total	Units
703 709	Joists 2 x 8 Joists 2 x 6	941 + 135 = 1076 fbm 833 fbm Cost per mfbm:	1.08 0.83	mfbm mfbm
	See Mat'ls. – 7 See Equip. – 7	Lumber = $150.00 Nails – 20 lb/mfbm @ .20 = $ 4.00 Material = $154.00		
	Table 7-1	Labor – 25 hr @ 7.14 = $178.50		
705	Bridging 1 x 4	29 fbm = Cost per mfbm:	0.03	mfbm
	See Mat'ls.– 7 See Equip. – 7	Lumber = $150.00 Nails – 45 lb/mfbm @ $.20 = $ 9.00 Material = $159.00		
		Labor – 50 hr/mfbm @ $7.14 = $357.00		
706	Studs 2 x 4	1,400 fbm = Cost per mfbm:	1.4	mfbm
	See Mat'ls.– 7 See Equip. – 7	Lumber = $150.00 Nails – 20 lb/mfbm @ $0.20 = $ 4.00 Material $154.00		
	Table 7-1	Labor – 20 hr @ $7.14 $142.80		
704 708	Headers – 2 x 8 Ribbons – 1 x 4	168 + 67 = 235 fbm 40 fbm Cost per mfbm:	0.24 0.04	mfbm mfbm
	See Mat'ls.– 7 See Equip. – 7	Lumber = $150.00 Nails – 20 lb/mfbm @ $0.20 = $ 4.00 Material $154.00		
		Labor 30 hr @ $7.14 = $214.20		
710	Rafters 2 x 6	952 + 73 + 80 + 93 = 1198 fbm Cost per mfbm:	1.20	mfbm
	See Mat'ls.– 7 See Equip. – 7	Lumber = $150.00 Nails – 20 lb/mfbm @ $0.20 = $ 4.00 Material $154.00		
	Table 7-1	Labor – 26 hours @ $7.14 = $185.64		

Estimate by _____ Date _____ Ckd. by _____ Date _____

Item No.	Identity & Cost Source	Computation of Unit Costs	Total	Units
707	Plates & caps	546 + 84 = 630 fbm Cost per mfbm:	0.70	mfbm
	See Mat'ls.- 7 See Equip. - 7	Lumber = $150.00 Nails - 20#/mfbm @ 0.20 = $ 4.00 Material $154.00		
	Table 7-1	Labor - 20 hr @ 7.14 = $142.80		
711	Termite shields	Cost per linear foot:	129	Lin ft
		Material - $0.25		
		Labor - $0.25		
712	Subfloor 1/2" plywood	1000 fbm = Cost msf:	1.	msf
	See Matl's. - 7	plywood msf. @ $130. = $130.00 Nails - 20 lb/msf @ 0.20 = $ 4.00 Material $134.00		
		Labor - 10 hr @ 7.14 = $ 71.40		
713	Sheathing 3/4" T & G for roof	1131 fbm = Cost per mfbm:	1.13	mfbm
	See Mat'ls. - 7 See Equip. - 7	Lumber @ $170/mfbm = $170.00 Nails - 20 lb/mfbm @ 0.20 = $ 4.00 Material $174.00		
		Labor 20 hr @ 7.14 = $142.80		

Estimate by _____ Date _____ Ckd. by _____ Date ____

Item No.	Identity & Cost Source	Computation of Unit Costs	Total	Units
714	Insulation in ceiling	4" mineral batts Cost per 100 sq ft: Material – 100 sq ft @ 0.09 = $ 9.00	9	100 sqft
	Table 7-1	Hrs to place 100 sq ft = 1.33 hr/1.33 sq ft = 1 hr Labor = 1 hr @ 7.14 = $ 7.14		
715	1/2" Celotex Undersiding	Cost per 100 sq ft: 1/2" Celotex @ $0.12/sq ft = $ 12.00 Nails – 2 lb @ 0.20 = $ 0.40 Material = $ 12.40 Labor – 1 hr @ $7.14 = $ 7.14	13.3	100 sqft
		FINISH CARPENTRY INTERIOR		
717	Hardwood floor 1 x 2 1/4	7.6 x 150 = 1140 fbm = See table 7-2	1.14	mfbm
718	Bldg paper (15 lb felt)	Cost per 1,000 fbm:		
	Table 7-2	(717) Flooring = $400.00 (717) Nails 10# x 10 @ 0.20 = $ 20.00 (718) 15 lb felt @ $0.75/100 sq ft = $ 7.50 Material $427.50		
	Table 7-2 Table 7-4	(717) Labor – 45 hr @ 7.14 = $321.30 (718) Labor 0.3 x 10 @ 7.14 = $ 21.42 Labor $342.72		
1100	1/2" Gypsum board on int. walls & ceiling (Drywall const) Table 11-7	9.5 + 21. = Cost per 100 sq ft: 1/2" Gypsum board @ 0.09/sq ft = $ 9.00 Tape for joints @ 0.01/sq ft = $ 1.00 Nails – 2 lb @ 0.20 = $ 0.40 Material $ 10.40	30.5	100 sqft

SUMMARIES & UNIT COSTS—Example 7-1 (continued)

Estimate by _____ Date _____ Ckd. by _____ Date _____

Item No.	Identity & Cost Source	Computation of Unit Costs	Total	Units
	Table 11-7	Nailing to studs and taping joints 2 hr Labor = $ 14.28		
1102	Ceramic tile floor 1" x 1" porcelain	32 sq ft = Cost per 100 sq ft:	.32	100 sqft
	Table 11-13	Tile (including bed) Material $74 x 2 = $148.00		
	Table 11-13	Labor (6.9 hr @ $14.00) 2 = $193.20		
1103	Ceramic tile walls 4 1/4" x 4 1/4" glazed	92 sq ft = Cost per 100 sq ft:	.92	100 sqft
	Table 11-13	Tile (including bed) Material $64 x 2 = $128.00		
		Labor ($11.0 x 14) 2 = $308.00		
1104	Vinyl tile 9" x 9" x .05"	88 sq ft = Cost per 100 sq ft:	.88	100 sqft
	Table 11-12	Tile (including paste) Material = $ 35.00		
		Labor 1 hr @ 13.85 = $ 13.85		
1016	Doors, int., birch	2/4 x 6/8 x 1 3/8 hollow core prehung Cost per door:	1	Each
	Table 10-4	Door $ 14.00 Prehung with frame hdw. & trim $ 50.00 Material $ 64.00		
		Labor 2.6 hr @ 7.14 = $ 18.56		
1017	Doors - Int., birch	2/0 x 6/8 x 1 3/8 hollow core prehung Cost per door:	5	Each
	Table 10-4	Door $ 13.00 Prehung with frame, hdw., & trim $ 50.00 Material $ 63.00		
		Labor 2.6 hr @ 7.14 = $ 18.56		

Estimate by _____ Date _____ Ckd. by _____ Date _____

Item No.	Identity & Cost Source	Computation of Unit Costs	Total	Units
1000	Windows 3' – 0" x 4' – 6" D. H. wood Table 10-7	Prehung with screens and weatherstripping Cost per window: Window – = $ 42.00 Weatherstripping = $ 10.00 Material $ 52.00 Labor 3 hr @ $7.14 = $ 21.42	14	Each
1010	Doors – Front solid core flush type pine Table 10-4	3/0 x 6/8 x 1 3/4 solid core, prehung Cost per door: Door – = $ 70.00 Prehung with frame, hdw. & trim = $ 60.00 Material $130.00 Labor 3 hr @ $7.14 = $ 21.42	1	Each
1011	Doors – Rear, solid core flush type pine Table 10-4	2/8 x 6/8 x 1 3/4 solid core, prehung Cost per door: Door – = $ 65.00 Prehung with frame, hdw. & trim = $ 60.00 Material $125.00 Labor 2.8 hr @ $7.14 = $ 20.00	1	Each
1015	Doors – Int., birch Table 10-4	2/8 x 6/8 x 1 3/8 hollow core prehung Cost per door: Door – = $ 16.00 Prehung with frame, hdw. & trim = $ 50.00 Material $ 66.00 Labor 2.6 hr @ $7.14 = $ 18.56	3	Each

Estimate by _____ Date _____ Ckd. by _____ Date _____

Item No.	Identity & Cost Source	Computation of Unit Costs	Total	Units
720	Interior trim Baseboard	Cost per 100 lin ft:	2	100 L.F.
		Baseboard @ 0.20/lin ft = $ 20.00 Nails 3 lb/100 lin ft @ 0.20 = $ ___.60 Material $ 20.60		
	Table 7-4	Labor - 6 hr @ $7.14 $ 42.84		
721	Shoe mold	Cost per 100 lin ft: Shoe mold @ $0.06/lin ft = $ 6.00 Nails = $ ___.60 Material $ 6.60	2	100 L.F.
	Table 7-4	Labor - 3 hr @ $7.14 $ 21.42		
722	Crown mold	Cost per 100 lin ft: Crown mold @ $0.12/lin ft = $ 12.00 Nails = $ ___.60 Material $ 12.60	2.5	100 L.F.
	Table 7-4	Labor - 4 hr @ $7.14 $ 28.56		
725	Kitchen Cabinets Base	Base cabinets - 24" deep x 36" high Cost per lin ft:	7.5	lin ft
		Material = $ 20.00		
		Labor - 1 hr @ $7.14 = $ 7.14		
726	Kitchen Cabinets Wall	Wall cabinets - 12" deep x 30" high Cost per lin ft:	13.0	lin ft
		Material $ 15.00		
		Labor - 1 hr @ $7.14 $ 7.14		

Estimate by _____ Date _____ Ckd. by _____ Date _____

Item No.	Identity & Cost Source	Computation of Unit Costs		Total	Units
730	Clothes closets	3 shelves and rods each Cost per closet:		4	Each
		Material	$ 10.00		
		Labor 1 1/2 hr @ $7.14	$ 10.71		
731	Closets- linen	5 shelves and rod Cost per closet:		1	Each
		Material	$ 15.00		
		Labor 2 hr @ $7.14	$ 14.28		
735	Hardware thresholds	3' long bronze saddles Cost each:		2	Each
		Material	$ 10.00		
		Labor 1 1/2 hr @ $7.14	$ 10.71		
		FINISH CARPENTRY - EXTERIOR			
410	Concrete entrances with steps	Cost per cu yd:		1	cu yd
		Concrete Form lumber Material	$ 20.00 $ 20.00 $ 40.00		
		Place and tamp dirt fill - 2 hr @ $7.14	= $ 14.28		
		Form labor - 4 hr @ $7.14	= $ 28.56		
		Place concrete 1 hour @ $7.14	= $ 7.14		
		Finish concrete 1 hour @ $7.70 Labor	= $ 7.70 $ 57.68		

Estimate by _____ Date _____ Ckd. by _____ Date _____

Item No.	Identity & Cost Source	Computation of Unit Costs	Total	Units
740	Bevel siding 6" siding	115 sq ft x $1.65 (See Table 7-3)	1.78	100 sqft
		Cost per 100 sq ft: Note - 100 sq ft up to 1" thick is 100 fbm		
	Table 7-3	Siding @ $275/mfbm = $ 27.50 Nails - 4 lb/100 sq ft @ $0.20 = $ 1.24 Material $ 28.74		
	Table 7-5	Labor - 5 hr/100 sq ft @ $7.14 = $ 35.70		

SUMMARIES & UNIT COSTS—Example 7-1 (continued)

Estimate by _____ Date _____ Ckd. by _____ Date _____

Item No.	Identity & Cost Source	Computation of Unit Costs	Total	Units
741	Building paper 15 lb felt	Cost per 100 sq ft	1.15	100 sqft
	See "ROOFING" Chapt. 9	Bldg. paper = \$ .75 Roofing nails - 1# @ \$0.35 = \$ .35 Material \$ 1.10		
		Labor 3 hr @ \$7.14 \$ 2.14		
745	3/8" Ext. plywood	Cost per msf:	.2	msf
	See Mat'ls - 7	Plywood = \$125.00 Nails - 20 lb/msf @ \$0.20 = \$ 4.00 Material \$129.00		
	Table 7-5	Labor - 8 hr @ \$7.14 \$ 57.12		
746	Fascias	(84 + 141 + 84 + 70) sq ft in these surfaces = 84 + 141/2 + 84/2 + 70/2 = Cost per msf:	.232	msf
		Wood = \$200.00 Nails 20/msf @ \$0.20 = \$ 4.00 Material \$204.00		
	Table 7-5	Labor 15 hr @ \$7.14 = \$107.10		
747	Molding at overhang	2" wide molding Cost per 100 lin ft:	150	100 L.F.
	See Molding and Trim Chapt. 7	Molding - 2" x \$0.04 x 100' = \$ 10.00 Nails 3 lb/100 lin ft @ \$0.20 = \$ .60 Material \$ 10.60		
		Labor 4 hr @ \$7.14 \$ 28.56		

Estimate by _____ Date _____ Ckd. by _____ Date _____

Item No.	Identity & Cost Source	Computation of Unit Costs	Total	Units
750	Shingle roof	Cost per square (100 sq ft):	11.5	100sqft
	See "Asphalt Shingles" Chapt. 9	Asphalt shingles - 240 lb/100 sq ft $\quad$ = $ 7.50 Nails 3 lb @ $0.35 $\quad$ = $ 1.05 Material $\quad\quad$ $ 8.25		
		Labor 3 hr @ $7.14 $\quad\quad$ $ 21.42		
751	Roofing felt 15 lb	Cost per square (100 sq ft):	11.5	100sqft
	See "ROOFING" Chapt. 9	Roofing felt $\quad\quad$ = $ 0.75 Nails $\quad\quad$ = $ 0.35 Material $\quad\quad$ $ 1.10		
		Labor 1 hr @ $7.14 $\quad\quad$ $ 7.14		
900	Flashing Chimney	16 oz copper Cost psf:	20	sq ft
	See "Flashing" Chapt. 9	Material cost $\quad\quad$ $ 1.25		
		Labor 2 hr @ $7.14 $\quad\quad$ $ 1.43		
901	Flashing - Ridge Strip	26 gage galv steel Cost per lin ft:	38	lin ft
	See "Flashing" Chapt. 9	Material cost $\quad\quad$ $ 0.30		
		Labor .04 hr @ $7.14 $\quad\quad$ $ 0.29		
902	Gutters - 5" Aluminum	Cost per lin ft:	76	lin ft
	See "Flashing" Chapt. 9	Material cost $\quad\quad$ $ 0.40		
		Labor .06 hr @ $7.14 $\quad\quad$ $ 0.43		

Estimate by _____ Date _____ Ckd. by _____ Date _____

Item No.	Identity & Cost Source	Computation of Unit Costs		Total	Units
903	Down spouts 5" Aluminum	Cost per lin ft:		40	lin ft
	See "Flashing" Chapt. 9	Material cost	$ 0.40		
		Labor .06 hr @ $7.14	$ 0.43		
1110	Varnish Hardwood floors	Cost psf:		700	sq ft
	Table 11-16	Material	$ 0.02		
		Labor	$ 0.05		
1111	Sanding floors	Cost psf:		700	sq ft
	Table 7-4	Labor .01 hr @ $7.14	$ 0.07		
1112	Painting Walls & ceiling latex primer & 2 coats	Cost psf:		3,050	sq ft
		Material	$ 0.04		
	Table 11-16	Labor	$ 0.12		
1113	Painting Kitchen Enamel primer & 2 coats	Cost psf:		330	sq ft
		Material	$ 0.05		
	Table 11-16	Labor	$ 0.13		
1114	Painting Bath Enamel primer & 2 coats	Cost psf:		180	sq ft
		Material	$ 0.05		
	Table 11-16	Labor	$ 0.13		

Estimate by ————————— Date ————————— Ckd. by ————————— Date —————————

Item No.	Identity & Cost Source	Computation of Unit Costs		Total	Units
1115	Painting windows Enamel primer & 2 coats	3.2 (See "Painting" Chapt. 11") x 175 sq ft = Cost psf:		560	sq ft
		Material	$ 0.06		
	Table 11-16	Labor	$ 0.13		
1116	Painting doors Enamel primer & 2 coats	2.6 (See "Painting" Chapt. 11) x 250 sq ft = Cost psf:		650	sq ft
		Material	$ 0.06		
	Table 11-16	Labor	$ 0.13		
1117	Painting Interior Trim Enamel primer & 2 coats	(200 + 200 + 250) x 2(See "Painting" Chapt. 11) = Cost psf:		1,300	sq ft
		Material	$ 0.06		
	Table 11-16	Labor	$ 0.18		
1118	Painting Woodwork Latex primer & 2 coats	420 + 400 Cost psf:		820	sq ft
		Material	$ 0.04		
	Table 11-16	Labor	$ 0.12		
1120	Painting Exterior Trim, oil base primer & 2 coats	(200 + 84 + 71 + 42 + 35 + 300) Cost psf:		732	sq ft
		Material	$ 0.06		
	Table 11-16	Labor	$ 0.25		
1119	Painting Wood siding Oil base primer & 2 coats	Cost psf:		115	sq ft
		Material	$ 0.07		
	Table 11-16	Labor	$ 0.15		

Estimate by _____ Date _____ Ckd. by _____ Date _____

Item No.	Identity & Cost Source	Computation of Unit Costs		Total	Units
1121	Painting equipment	For brushes, ladders, etc., for small jobs allow:			
		Equipment	$ 40.00		
765	Vents in foundation	6" x 8" cast iron Cost each:		10	Each
		Material	$ 3.00		
		Labor	$ 1.00		
1500	Heating and air conditioning	Assume 12% of $15,000	$1800.00	1	L. S.
1510	5' P.E. tub	(See Table-Chapt. 15-for cost data on plumbing)		1	Each
		Material	$ 300.00		
		Labor 7.5 hr @ $15.45	$ 116.00		
1511	Lavatory	Material	$ 160.00	1	Each
		Labor 6 hr @ $15.45	$ 93.00		
1512	W. C.	Material	$ 225.00	1	Each
		Labor 8.5 hr @ $15.45	$ 131.00		
1513	Medicine cabinet	Material	$ 35.00	1	Each
		Labor 1 hr @ $15.45	$ 16.00		
1514	Water heater	Material	$ 185.00	1	Each
		Labor 5 hr @ $15.45	$ 77.00		

SUMMARIES & UNIT COSTS—Example 7-1 (continued)

Estimate by _____ Date _____ Ckd. by _____ Date _____

Item No.	Identity & Cost Source	Computation of Unit Costs		Total	Units
1515	Kitchen sink	Material	$ 160.00	1	Each
		Labor 7.5 hr @ $15.45	$ 116.00		
1516	Water – service from main 3/4" copper	Material 75' x $1.20	$ 90.00		
		Labor 75' x .1 hr @ $15.45	$ 116.00		
1517	4" C.I. Sewer	Material 75' x $2.40	$ 180.00		
		Labor 75' x .3 hr @ $15.45	$ 348.00		
1520	Electrical	Assume 7.5% of $15,000	$1125.00	1,125	L. S.

DIRECT COSTS—Example 7-1

Estimate by _____ Date _____ Ckd. by _____ Date _____

Item No.	Identity & Location	Quantity		Unit Cost Each			Total Cost Each			Total Cost
		No.	Unit	Equip.	Mat'l.	Labor	Equip.	Mat'l.	Labor	
300	Clearing & grubbing									$ 300
301	Excavating								$ 150	150
400	Conc. foot	3.8	cu yd	$ 20.00	$ 4.00			$ 76	15	91
410	Conc. ent.	1.0	cu yd		40.00	57.68		40	58	98
600	8" Block	2.4	100		37.70	80.30		90	193	283
601	8" Block	.1	100		64.70	80.30		6	8	14
602	4" Block	.1	100		36.80	61.03		4	6	10
610	Brick	.6	1000	$ 13.00	75.91	189.80	$ 8.00	46	114	168
611	Met. ties	818	Each		.02			16		16
612	Brick	1.5	1000	19.00	75.91	379.60	29.00	114	569	712
700	2 x 10 Lum.	.18	mfbm		164.00	178.50		30	32	62
701	Equip.						90.00			90
702	Ledgers	.03	mfbm		156.00	285.60		5	9	14
703	Joists 2 x 8	1.08	mfbm		154.00	178.50		166	193	359
704	Headers	.24	mfbm		154.00	214.20		37	51	88
705	Bridging	.03	mfbm		159.00	357.00		5	11	16
706	Studs	1.4	mfbm		154.00	142.80		216	200	416
707	Pls & caps	.7	mfbm		154.00	142.80		108	100	208
708	Ribbons	.04	mfbm		154.00	214.20		6	9	15
709	Joists 2 x 6	.83	mfbm		154.00	178.50		128	148	276
710	Rafters	1.20	mfbm		154.00	185.64		185	223	408
711	Termite sh.	129	Lin ft		.25	.25		32	32	64
712	Subfloor	1.	msf		134.00	71.40		134	71	205
713	Sheathing	1.13	mfbm		174.00	142.80		197	161	358
714	Ceiling ins.	9.	100 sf		9.00	7.14		81	64	145
715	Celotex	13.3	100 sf		12.40	7.14		165	95	260
717 718	Floor & Bldg. paper	1.14	mfbm		427.50	342.72		487	391	878
720	Baseboard	2.	100 Lf		20.60	42.84		41	86	127
721	Shoe mold	2.	100 Lf		6.60	21.42		13	43	56
722	Crown mold	2.5	100 Lf		12.60	28.56		32	71	103
725	K. cab. base	7.5	Lin ft		20.00	7.14		150	54	204
726	K. cab. wall	13.0	Lin ft		15.00	7.14		195	93	288
730	Clo. closet	4.	Each		10.00	10.71		40	43	83
731	Lin. closet	1.	Each		15.00	14.28		15	14	29
735	Threshold	2.	Each		10.00	10.71		10	11	21
							$127	$2,870	$3,318	$6,615

Estimate by ─────────────── Date ─────────── Ckd. by ─────────── Date ──────────

Item No.	Identity & Location	Quantity		Unit Cost Each			Total Cost Each			Total Cost
		No.	Unit	Equip.	Mat'l.	Labor	Equip.	Mat'l.	Labor	
740	Siding	1.78	100 sf		$ 28.74	$ 35.70		$ 51	$ 64	$ 115
741	Bldg. paper	1.15	100 sf		1.10	2.14		1	2	3
745	Ext. plywood	.2	msf		129.00	57.12		26	11	37
746	Fascias	.232	msf		204.00	107.10		47	25	72
747	Molding	1.5	100 Lf		10.60	28.56		16	43	59
750	Shingles	11.5	100 sf		8.25	21.42		95	246	341
751	Roof felt	11.5	100 sf		1.10	7.14		13	82	95
765	Fdn. vents	10.	Each		3.00	1.00		30	10	40
900	Chim. flash	20.	sf		1.25	1.43		25	29	54
901	Ridge strip	38.	Lin ft		.30	.29		11	11	22
902	Gutters	76.	Lin ft		.40	.43		30	33	63
903	Dn. spouts	40.	Lin ft		.40	.43		16	17	33
1000	Windows	14.	Each		52.00	21.42		728	300	1,028
1010	Doors - Fr.	1.	Each		130.00	21.42		130	21	151
1011	Doors - Rr.	1.	Each		125.00	20.00		125	20	145
1015	Doors - Int.	3.	Each		66.00	18.56		198	56	254
1016	Doors - Int.	1.	Each		64.00	18.56		64	19	83
1017	Doors - Int.	5.	Each		63.00	18.56		63	18	81
1100	Drywall	30.5	100 sf		10.40	14.28		317	436	753
1102	Cer. tile	.32	100 sf		148.00	193.20		47	62	109
1103	Cer. tile	.92	100 sf		128.00	308.00		118	283	401
1104	Vinyl tile	.88	100 sf		35.00	13.85		31	12	43
1110	Varn. floor	700	sf		.02	.03		14	21	35
1111	Sand. floor	700	sf			.07			49	49
1112	Paint w & c	3050	sf		.04	.12		122	366	488
1113	Paint Kitchen	330	sf		.05	.13		17	43	60
1114	Paint Bath	180	sf		.05	.13		9	23	32
1115	Paint Windows	560	sf		.06	.13		34	73	107
1116	Paint Doors	650	sf		.06	.13		39	85	124
1117	" Int. trim	1300	sf		.06	.18		78	234	312
1118	" Woodwork	820	sf		.04	.12		33	98	131
1119	" Siding	115	sf		.07	.15		8	17	25
1120	" Ext. trim	732	sf		.06	.25		44	183	227
1121	" Equip.	1	Each	40			40			40
1500	Heating	1	Each							1,800
							$ 40	$2,580	$2,992	$7,412

DIRECT COSTS—Example 7-1 (continued)

Estimate by _____ Date _____ Ckd. by _____ Date _____

Item No.	Identity & Location	Quantity		Unit Cost Each			Total Cost Each			Total Cost
		No.	Unit	Equip.	Mat'l.	Labor	Equip.	Mat'l.	Labor	
1510	P.E. tub	1	Each		300.00	116.00		$ 300	$ 116	$ 416
1511	Lav.	1	Each		160.00	93.00		160	93	253
1512	W.C.	1	Each		225.00	131.00		225	131	356
1513	Med. cab.	1	Each		35.00	16.00		35	16	51
1514	Water heater	1	Each		185.00	77.00		185	77	262
1515	Kitch. sink	1	Each		160.00	116.00		160	116	276
1516	Water serv.	75	Lin ft		1.20	1.55		90	116	206
1517	C.I. Sewer	75	Lin ft		2.40	4.64		180	348	528
1520	Electrical	1	Each							1,125
				Totals this sheet				1,335	1,013	3,473
				Totals other sheet			127	2,870	3,318	6,615
				Totals other sheet			40	2,580	2,992	7,412
				Total Direct Costs			$ 167	$6,785	$7,323	$17,500

OVERHEAD & PROFIT—Example 7-1

Estimate by _____ Date _____ Ckd. by _____ Date _____

Item No.	Class of Expense	Computations of Overhead Expense	Total Cost
1600	Gen. overhead (% direct cost)	1% x $17,500	175
	Job overhead	Assume 3 months for job	
1700	Int. on operating capital	10% x $17,500/2 x .09 (int.) x 3/12	20
1701	Superintendent's salary	1/6 for job = 1/6 x 10,000 x 3/12	416
1702	Supt. pickup truck - rental	1/6 for job = 100 x 1/6 x 3 months	50
	do. operating cost	1/6 for job - 3000 x 1/6 @ 10¢ per mile	50
1703	Job trucks - rental		
	do. operating cost		
	do. wages - driver's		
1704	Lifting equipment		
	do. operating cost		
	do. wages - operator's		
1705	Job office - rental		
	do. salaries		
	do. supplies		
1706	Utilities & connections		40
1707	Social Security	5.85% of $7,323 x 90%	386
1708	Workmen's Compensation	2.44% of $7,323	179
1709	Pub. Lia. & Prop. Damage	0.2% of $7,323	15
1710	Fed. & State Unemp. Ins.	3.2% x 45% x $7,323	105
1711	Patents & royalties		
1712	Barricades		25
1713	Temporary toilets		25
1714	Cut and patch for trades		119
1715	Permits		50
1716	Protection adjacent prop.		
1717	Final cleanup		50
	Subtotal of computed overhead		$ 1,541
1718	Contingencies (% of computed overhead) 5% of $1,541		$ 77
	Total overhead		$ 1,618
	Direct cost from Direct Cost sheet		$ 17,500
	Subtotal (Total overhead plus direct cost)		$ 19,118
	Profit 15% of $19,118		$ 2,868
	Total cost		$ 21,986
	Performance bond $21,986 x 1/1000 x $10		$ 220
	Total amount of bid		$ 22,206
	22,206/2483 x 36.83 = 24.30 psf		

HOMEWORK EXAMPLE 7-1

Make a complete estimate for the house plans in Figure 7-7 (also prepared by architect Robert B. Lyons of Raleigh, North Carolina) and show cost per square foot. Allow $350 for clearing & grubbing.

[The house plans (Figure 7-7) appear on pp. 190-197.]

FIGURE 7-7
Blueprints for a House

190

F.S. TRIM — FRONT DOOR

HEATING NOTES.
FURNACE TO BE DELCO MODEL OPC-100-LD RATED AT 100,000 B.TU/HR.
REGISTERS TO BE NATIONAL P-2 OR EQUAL-GRILLES-N4C-190 SERIES.
GAS SUPPLY TO TANK. DAY & NIGHT CLOCK THERMOSTAT
TOTAL HEAT LOSS- 90,585 B.T.U.

HEAT LOSS PER ROOM.
LIVING ROOM   19,140 BTU
KITCHEN   15,870 "
FAMILY ROOM   19,140 "
HALL   1,380 "
BATH RM 1   5,168 "
BATH RM 2   4,554 "
BED ROOM 1   13,351 "
BED ROOM 2   6,270 "
BED ROOM 3   10,212 "

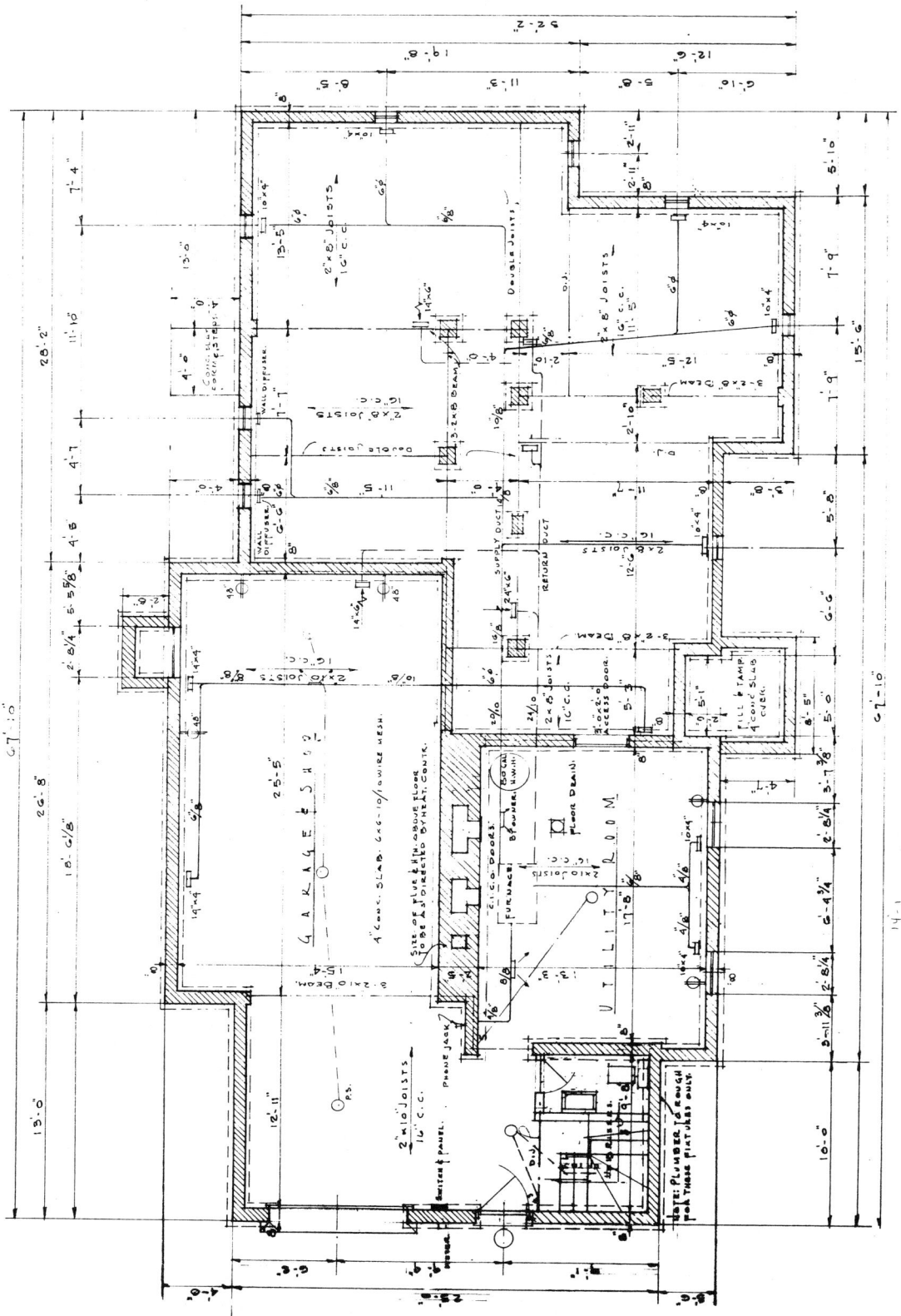

BASEMENT & FOUNDATION PLAN
SCALE-¼"=1'-0

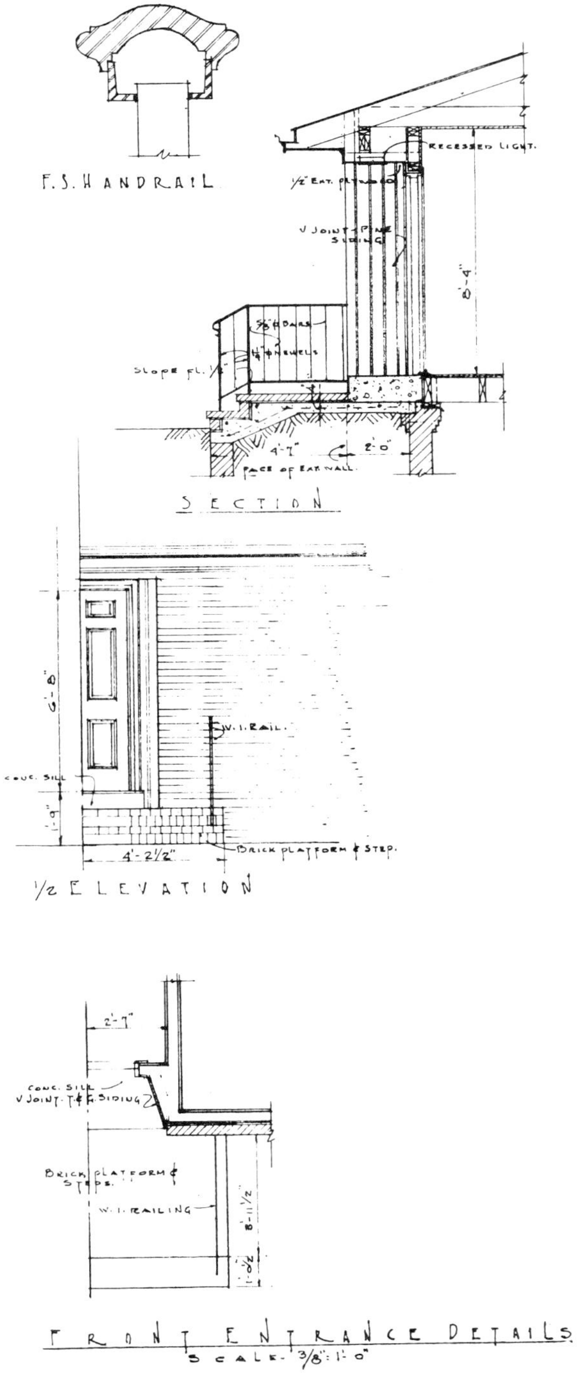

F.S. HANDRAIL
RECESSED LIGHT
1/2" EXT. PLYWOOD
V JOINT PINE SIDING
8'-4"
5/8" BARS
1 1/4" NEWELS
SLOPE FL. 1/4"
4'-7"
2'-0"
FACE OF EXT. WALL
SECTION
6'-8"
W.I. RAIL
CONC. SILL
9"
4'-2 1/2"
BRICK PLATFORM & STEP.
1/2 ELEVATION
2'-7"
CONC. SILL
V JOINT T&G SIDING
BRICK PLATFORM & STEPS.
W.I. RAILING
8'-11 1/2"
FRONT ENTRANCE DETAILS
SCALE 3/8" = 1'-0"

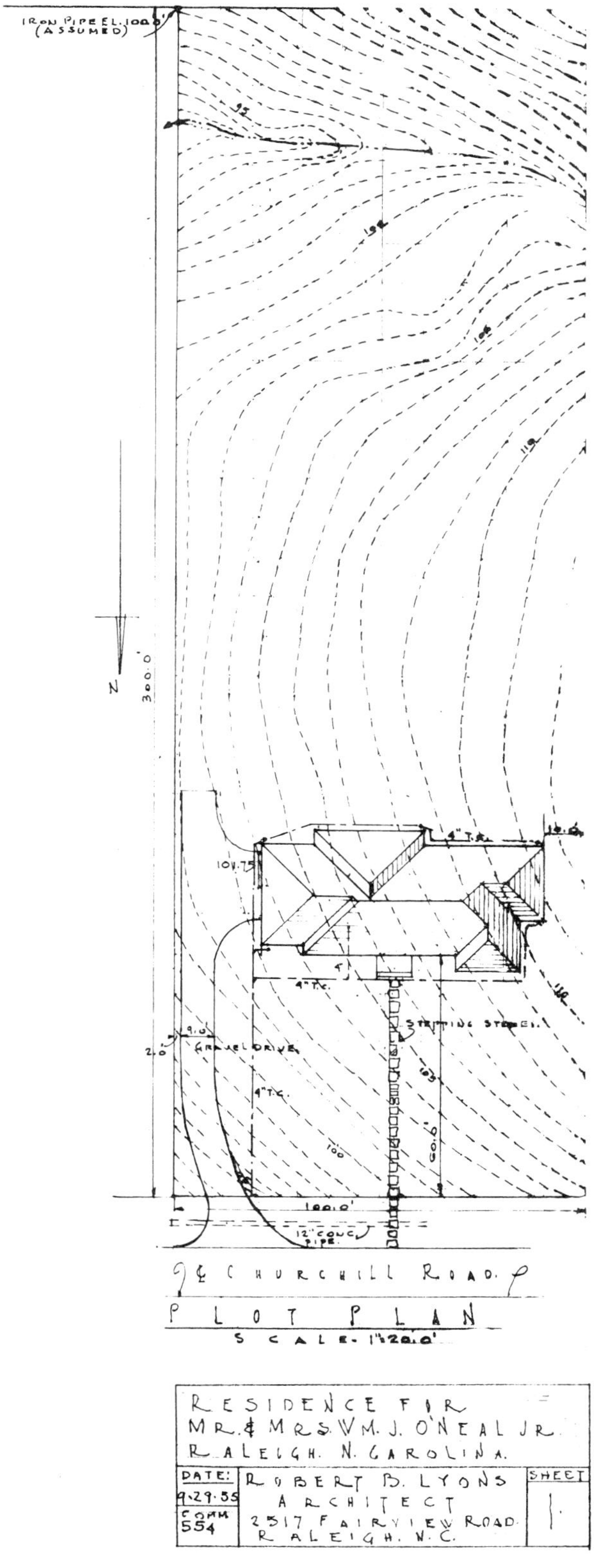

IRON PIPE EL. 100.0' (ASSUMED)
N
300.0'
104.75
4" T.C.
STEPPING STONES.
GRAVEL DRIVE
60.0'
100.0'
12" CONC. PIPE
℄ CHURCHILL ROAD ℄
PLOT PLAN
SCALE 1" = 20.0'
RESIDENCE FOR
MR. & MRS. WM. J. O'NEAL JR.
RALEIGH. N. CAROLINA.
DATE: 9.29.55
COMM 554
ROBERT B. LYONS
ARCHITECT
2517 FAIRVIEW ROAD
RALEIGH. N.C.
SHEET 1.

FIGURE 7-7 (continued)

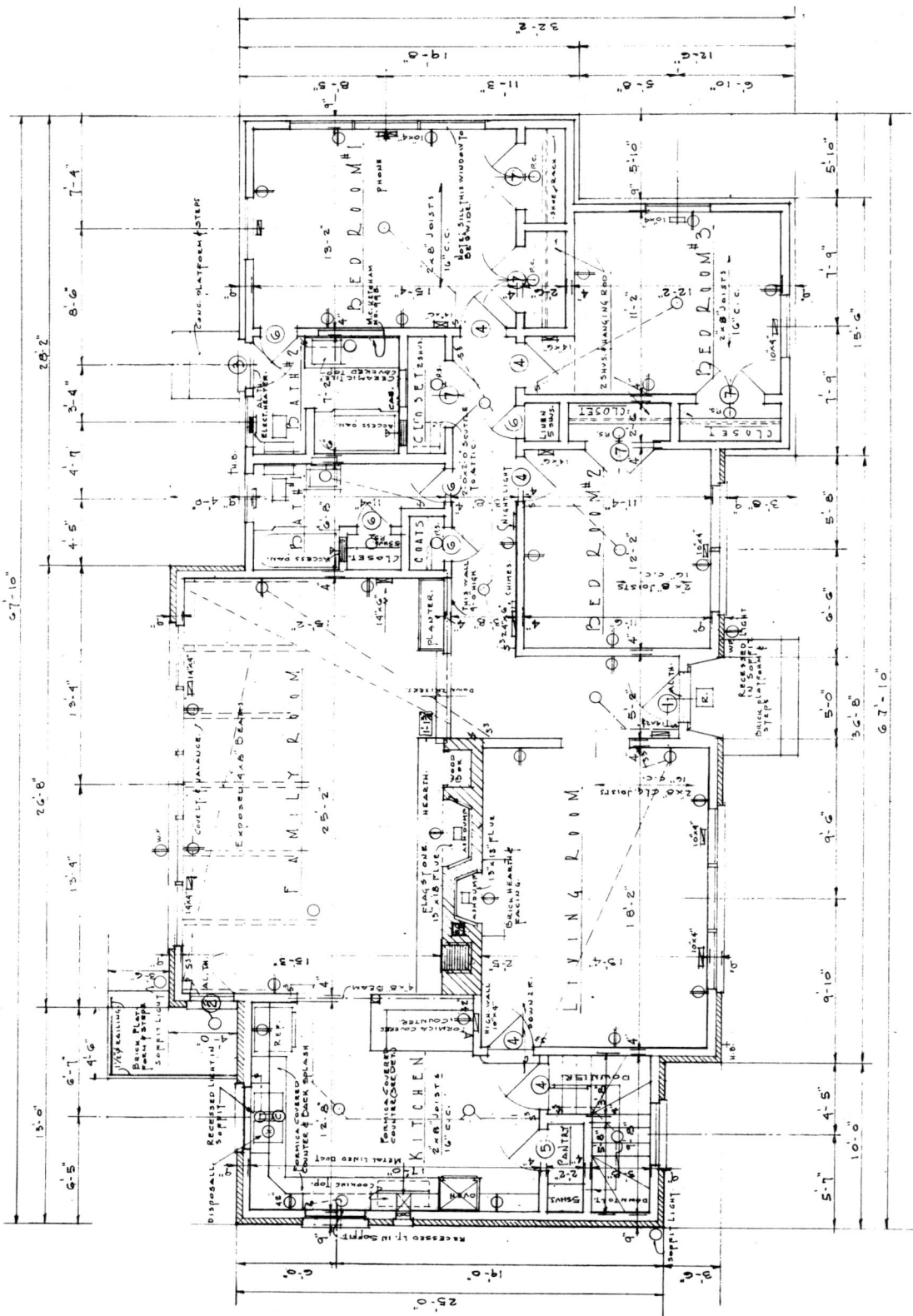

FIRST FLOOR PLAN
SCALE: 1/4" = 1'-0"

FIGURE 7-7 (continued)

192

DOOR SCHEDULE

MK	SIZE	TYPE	DESCRIPTION
1	3/0x6/8x1¾"	WOOD	FLUSH PANEL-SO.CO. (GUM).
2	3/0x6/8x1¾	"	
3	2/8x6/8x1¾	"	
4	2/8x6/8x1⅜	"	FL.PAN.HOLLOW CORE
5	2/6x6/8x1⅜	"	
6	2/0x6/8x1⅜	"	LOUVERED DOORS
7	1PR.2/0x6/8x1⅜	"	FLUSH PAN. SO.CORE.
8	3/0x6/8x1¾	"	FLUSH PAN. SO.CORE.
9	8'-0 x6/8x1⅜	"	OVERHEAD GARAGE DOOR.

WINDOW SCHEDULE

MK	SIZE	TYPE
A	3'-6x4'-8"	FLEXIVENT - ANDERSEN
B	3'-6x2'-8	2 SECT. AWNING TYPE WD GATE CITY
C	3'-6"x4'4	"
D	3'0"x2'-8"	2 "
F	2'-8"x5'-4"	3 "
G	2'-4"x1'-8"	BASEMENT SASH-ANDERSEN.

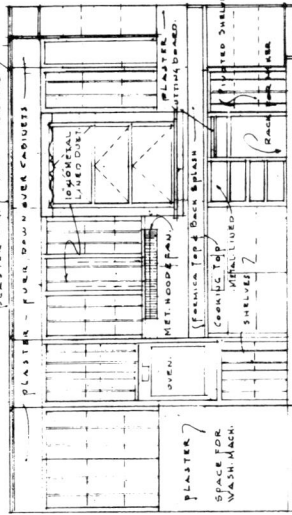

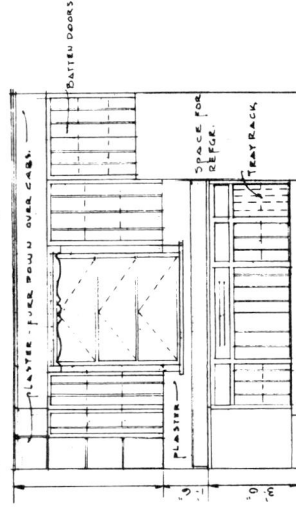

ROOM FINISHS SCHEDULE

ROOM	FLOOR	BASE	WAINSCOT	WALLS	CEILING	REMARKS
ENTRANCE HALL	WOOD	WOOD		PLASTER	PLASTER	
LIVING ROOM	"			PLASTER	"	
KITCHEN	VINYL TILE					
FAMILY ROOM	"			PANELING		
HALL	WOOD	WOOD		PLASTER	PLASTER	
BED ROOMS	WOOD	WOOD		"	"	
BATH ROOMS	VINYL TILE	WOOD				TILE WAINSCOT-6'-0 HIGH AROUND TUB

PLUMBING FIXTURE SCHEDULE

BATH RM. No.1
BATH TUB · 5'-0" MASTER PEMBROKE P 2227 I
WATER CLOSET-COMPACT ELONGATED F2040A
LAVATORY - LEDGEWOOD · 22x19" - P 4100A

BATH RM. No.2
BATH TUB · 5'-0" MASTER PEMBROKE P 2227 I
WATER CLOSET-COMPACT ELONGATED F2040A
LAVATORY

KITCHEN SINK
2 COMPARTMENT SINK · 42x20 P 10577 D
D-B1G ST FITTINGS
NOTE: PLUMBER TO

INSTALL DISPOSALL-SUPPLIED BY OWNER.

PLUMBING ACCESSORIES.

BATH RM. No.1 & No.2.
TOWEL BARS: 2 - 24" NO.G94
SOAP-TUMBLER&T.B.HOLDER NO.336
TOILET PAPER HOLDER- NO.G71
SOAP HOLDER - NO.G22
GRAB BAR - 16" NO.G60
ROBE HOOK - NO.G81.

NOTE: NUMBERS FOR PLUMBING ACCESSORIES ARE TAKEN FROM HALL MACK CATALOGUE.

DET. BATH 2 MN No.1.
SCALE 3/8"=1'-0

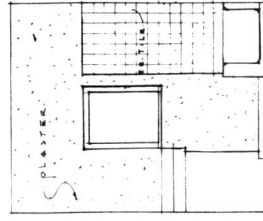

RESIDENCE FOR
MR. & MRS.WM.J.ONEAL-JR.
RALEIGH, N.C.
ROBERT D LYONS
ARCHITECT
2517 FAIRVIEW ROAD
RALEIGH N.C.

| DATE 9-29-55 | SHEET 2 |
| COMM 554 | |

2 of 3 FIRST FLOOR.

KITCHEN DETAILS
SCALE 3/8"=1'-0"

FIGURE 7-7 (continued)

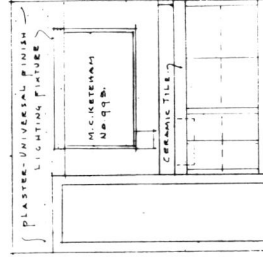

193

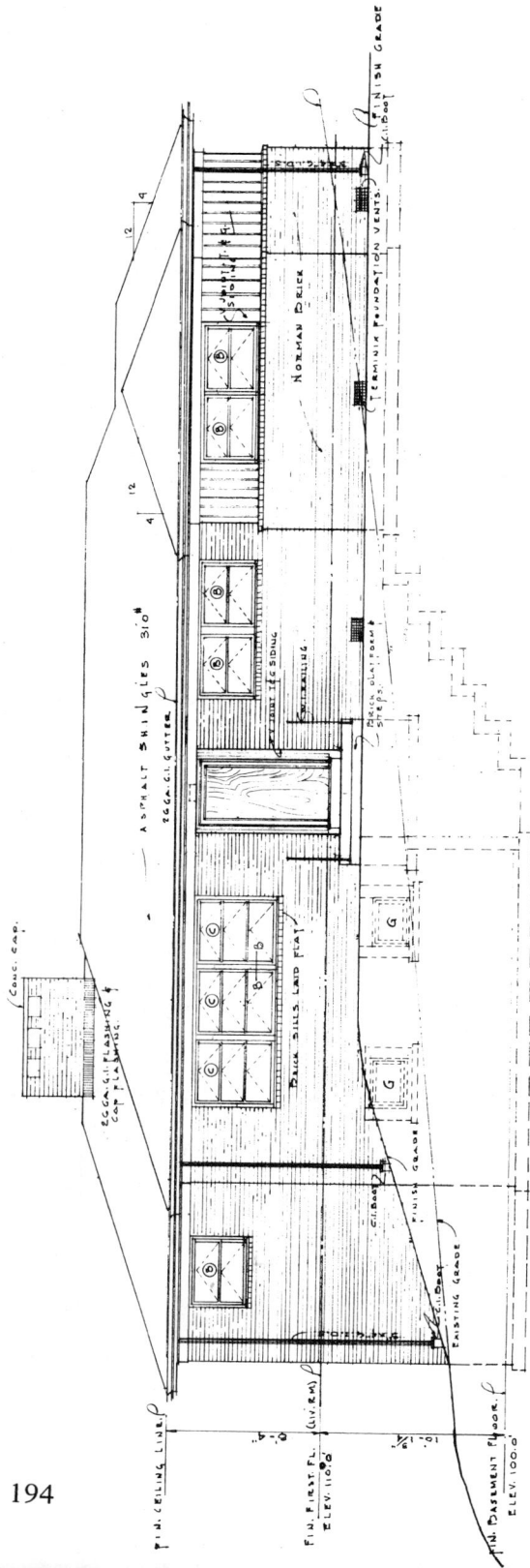

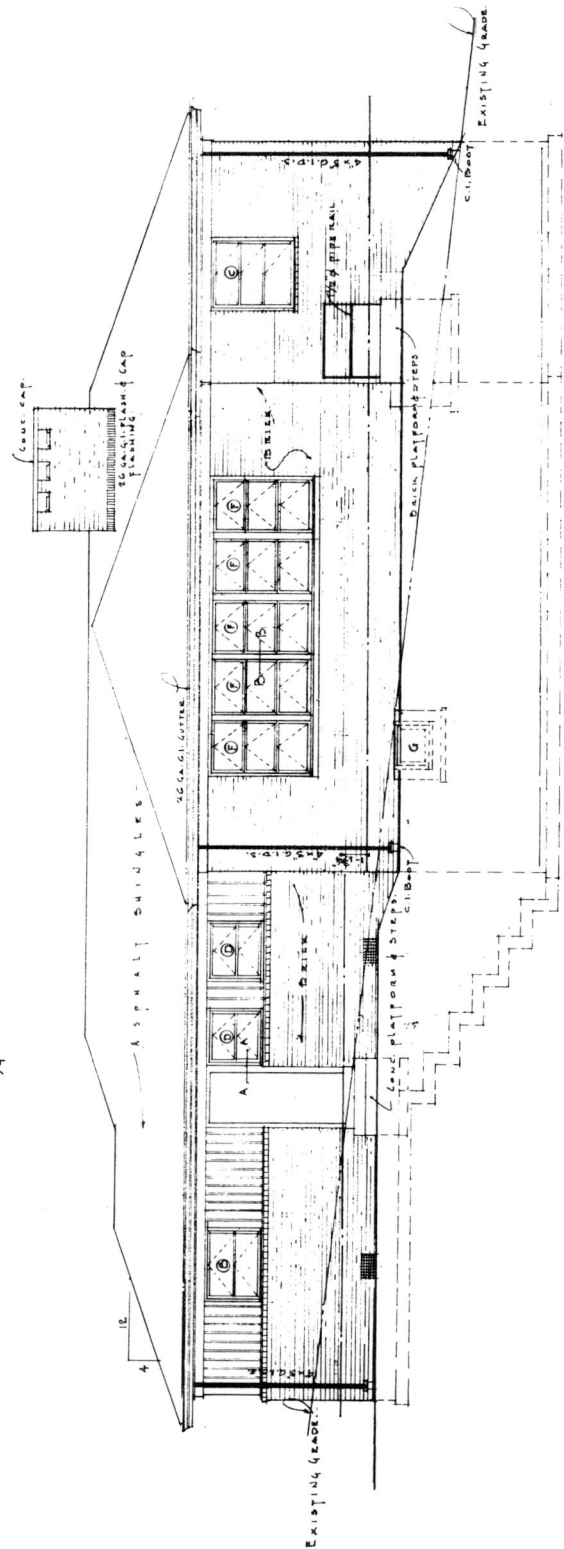

NORTH ELEVATION

SCALE: 1/4" = 1'-0"

SOUTH ELEVATION

SCALE: 1/4" = 1'-0"

194

FIGURE 7-7 (continued)

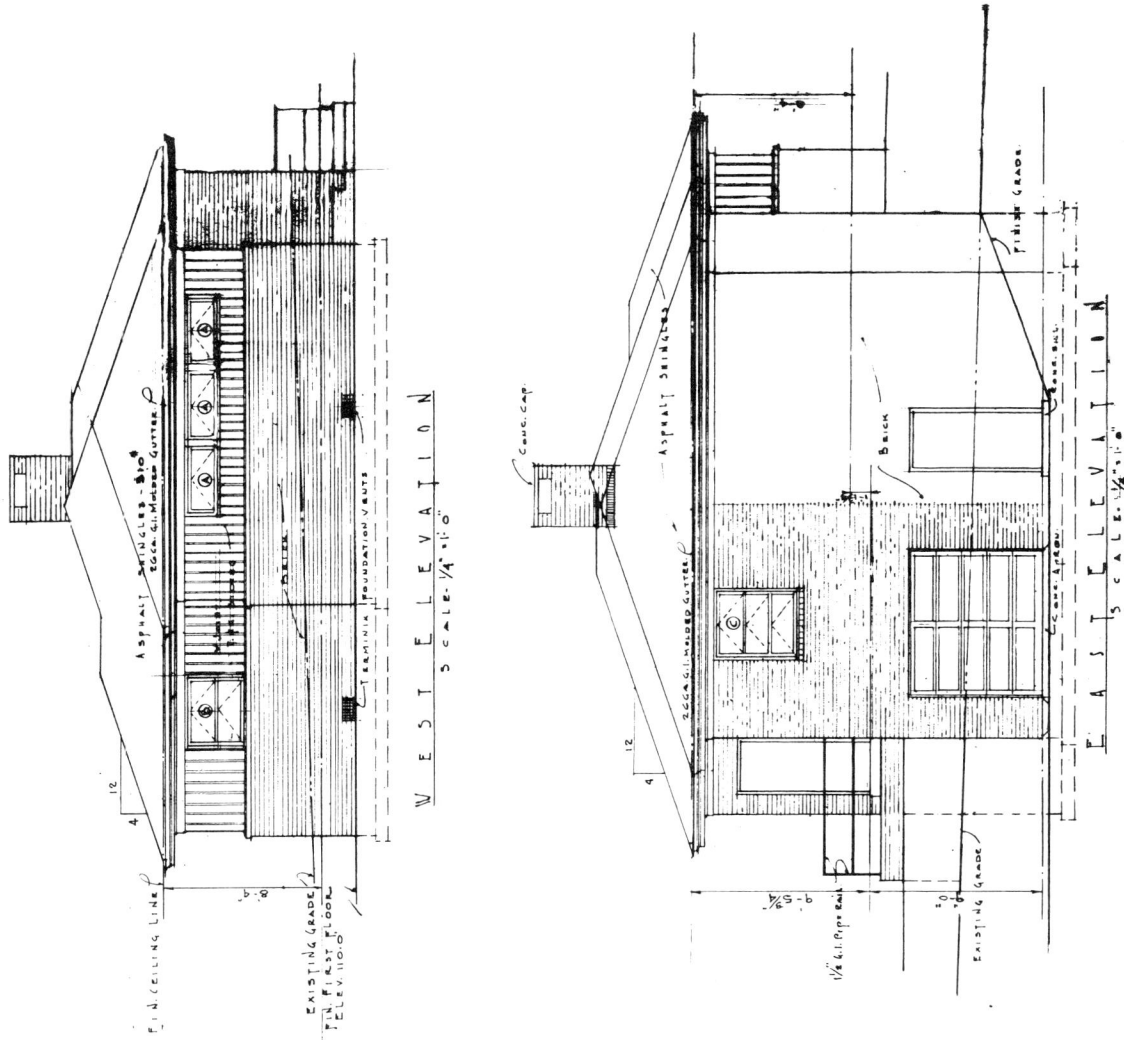

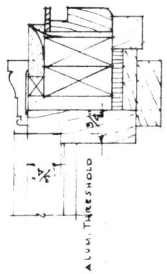

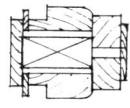

FIGURE 7-7 (continued)

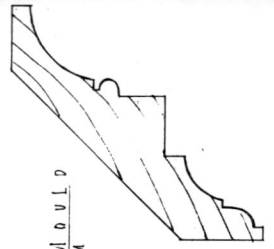

F.S. CROWN MOULD
FOR LIVING ROOM

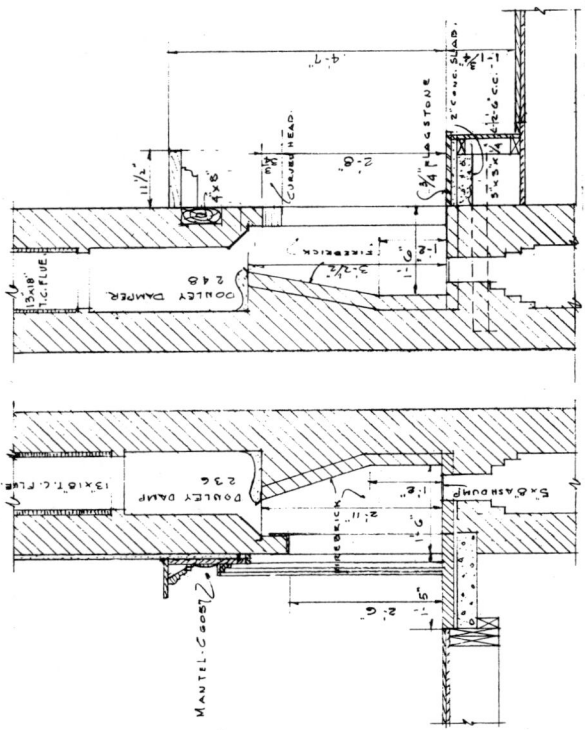

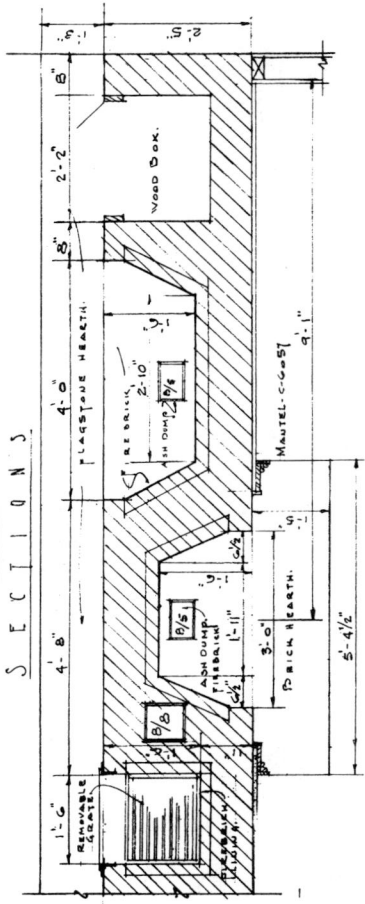

SECTIONS

FIREPLACE DETAILS
SCALE- 3/4"=1'-0"

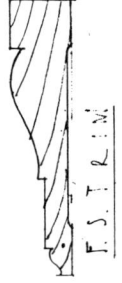

F.S. TRIM

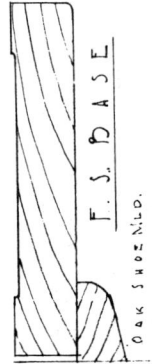

F.S. BASE
OAK SHOE MLD.

FIGURE 7-7 (continued)

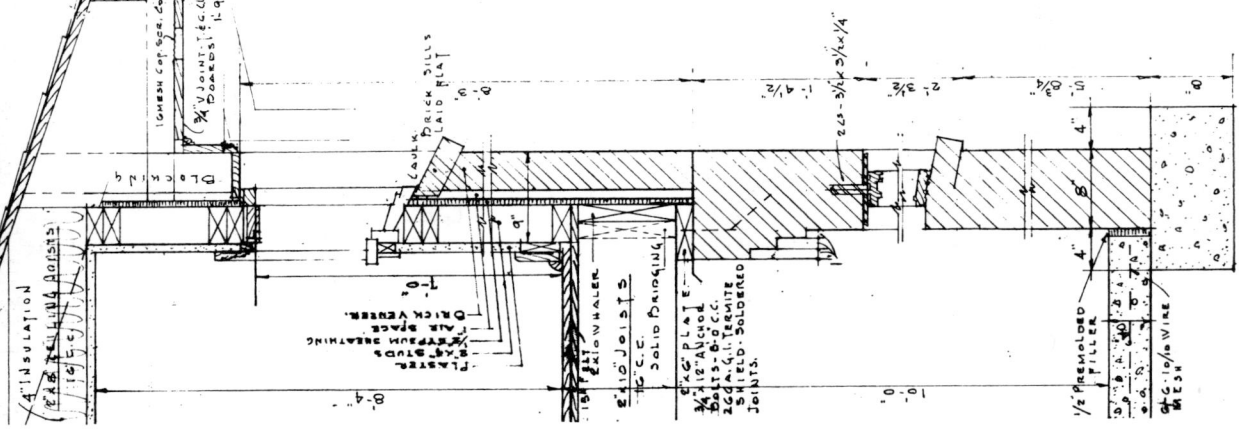

WALL SECTION
SCALE-1 1/2"=1'-0"

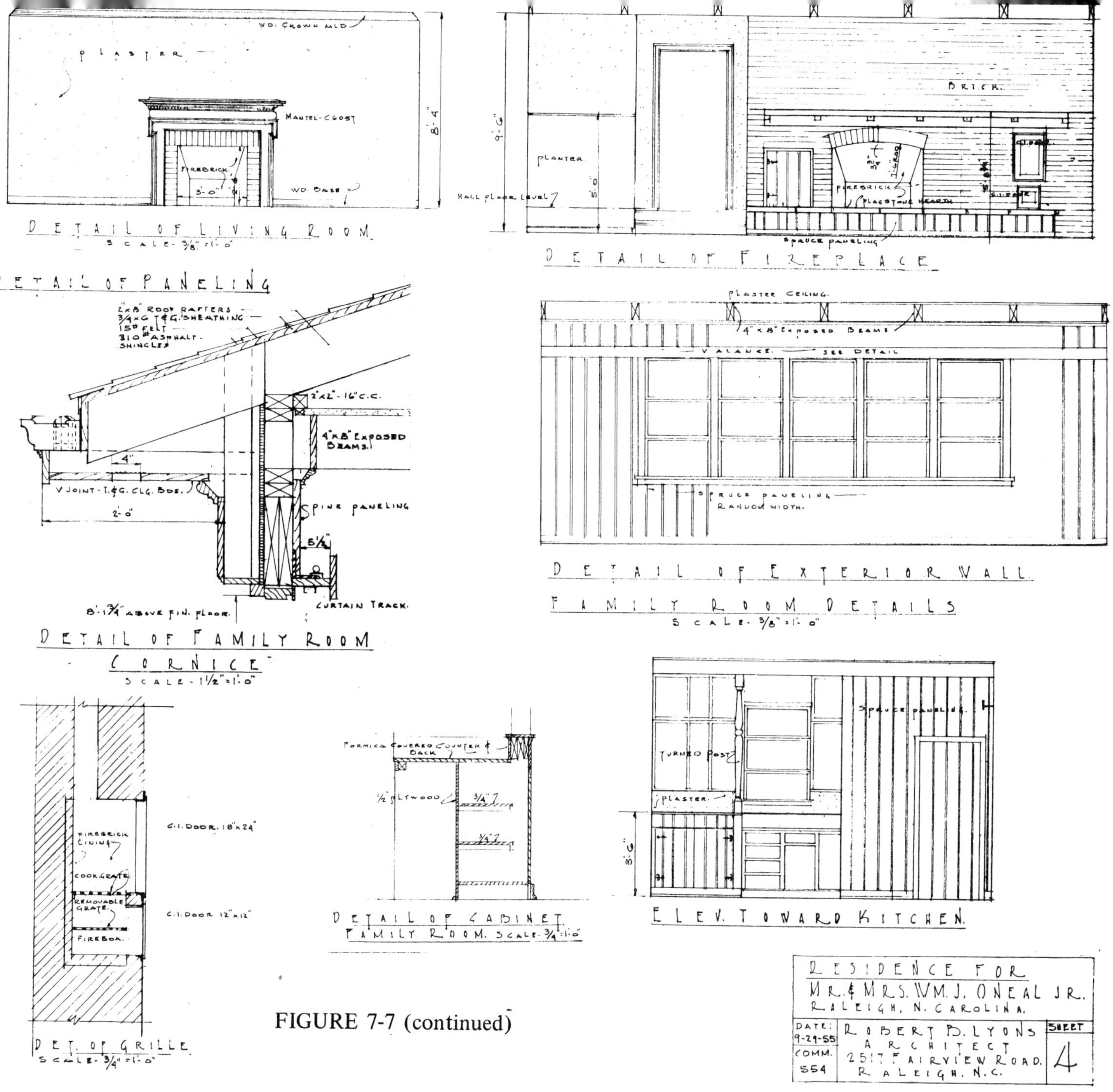

DETAIL OF FIREPLACE

DETAIL OF EXTERIOR WALL

FAMILY ROOM DETAILS
SCALE: 3/8"=1'-0"

ELEV. TOWARD KITCHEN

DETAIL OF LIVING ROOM
SCALE: 3/8"=1'-0"

DETAIL OF PANELING

DETAIL OF CABINET
FAMILY ROOM. SCALE 3/4"=1'-0"

DETAIL OF FAMILY ROOM
CORNICE
SCALE: 1 1/2"=1'-0"

DETAIL OF GRILLE
SCALE: 3/4"=1'-0"

RESIDENCE FOR
MR. & MRS. WM. J. O'NEAL JR.
RALEIGH, N. CAROLINA.
ROBERT P. LYONS
ARCHITECT
2517 FAIRVIEW ROAD,
RALEIGH, N.C.

| DATE: 9-24-55 | SHEET 4 |
| COMM. 554 | |

FIGURE 7-7 (continued)

197

CHAPTER EIGHT

Laminated Construction

Laminated construction is being used more and more. One great advantage of this type of construction is that all of the knots and other imperfections that reduce the strength of timber can be cut out of the length of each piece being prepared to go into the unit and the ends of the good material joined by means of a glued finger joint, thus leaving only the high-strength timber in the completed member.

METHOD OF CONSTRUCTION

Lamination consists of several pieces of wood being glued together to form a single piece: for instance, 24 pieces of 5 1/2″ × 3/4″ assembled as a unit would produce a beam 5 1/2″ wide × 18″ deep that could be assembled straight or bent into a curve to form an arch. The depth of the arch could be varied, say, from 18" at the center to 9" at the end by omitting one piece of lamination at a time from the center toward the ends and then planing the ends smooth to conform to the curve.

Advantages

By lamination wood members can be made continuous for long lengths up to a hundred feet or more, which would be impossible to accomplish in the making of an arch from a single piece of wood; even the making of a straight beam from one piece of wood of any considerable length would be prohibitive especially if no defects could be allowed. Laminated members contain only the perfect wood. The glued joints are as strong as or stronger than the original wood and when a waterproof glue is used the laminated units are safe for exposed work.

These members come finished with connection plates or other fastening devices installed ready for erection in much the same manner that fabricated steel is delivered to the job site. But, unlike steel, in most cases the wood members do not have to be encased but are intended to be seen. They can even come varnished requiring no finish. Widely spaced laminated beams with thick, tongue-and-grove, wood decking can actually, in many cases, be cheaper than bar joists with their hung ceilings which are required for fire protection or cosmetic reasons.

Where Used

Because of its beauty and reasonable cost in place laminated construction is frequently employed in arches and beams for churches, auditoriums, field houses, bowling alley centers, filling stations, and even in automobile show rooms.

Computing Costs

The cost of laminated members is controlled by many factors, some of which are: lumber, glue, labor, equipment, handling and shipping, and erection. None of these factors can be arrived at in the same manner that regular wood construction is estimated.

The lumber must be of very high quality with all blemishes removed, making this material cost higher than for ordinary work.

The cost of the glue varies with the thickness of the laminated pieces; for instance, if 1-in. instead of 2-in. material is used it will take twice as much glue, which is a very expensive item (especially when of the waterproof variety).

The labor varies with the complexity of the work, the amount of bending, and the size and length of the pieces. Very specialized equipment, much labor, and much time are required for arches, while straight beams require less equipment, time, and labor.

Time is an important factor since depreciation on the complex plant and equipment is very high.

Erection

Erection is one phase of the work in which this product excels. The members can be as easily erected as structural steel since they come with all of the fastenings built into the wood. It is even easier than steel because the bolts are much fewer for the wood and the members usually come varnished and finished just as they will appear in the completed structure. Even when they must be given a finished coat after erection the wood is already prepared for the varnish or paint. Also, no encasement is required as is usually the case with steel members and no curing and rubbing as with concrete members.

Principal Use

The principal use of laminated construction is for roof framing and decking. The beams or arches are usually spaced in the range of 8 to 24 feet on centers and covered with heavy timber decking 2-, 3-, 4-, or more in. thick. The decking is not laminated; it is, however, made tongue-and-groove with beveled edges and decorative patterns. Also, like the beams and arches it is usually varnished and requires no further finishing.

Approximate Cost Tables as of August 1970

TABLE 8-1
Supporting Members—Approximate Cost
per fbm fob Plant

Type of members	Span	Spacing	Cost per fbm
Straight laminated beams	30′ to 60′ clear	8′ to 20′ cts.	$0.50
Three-hinged "V" arches	40′ to 100′ clear	12′ to 20′ cts.	0.55
Segmented barrel arches	80′ to 200′ clear	16′ to 24′ cts.	0.50

TABLE 8-2
Heavy Timber Decking Approximate Cost per
1,000 fbm fob Plant

Thickness	Width	Conversion factor*	Cost per 1,000 fbm
2″	6″	2.4	$180
3″	6″	3.43	$200
4″	6″	4.58	$200

*To convert sq ft to fbm due to tongue-and-groove effect.

TABLE 8-3
Approximate Trucking Cost per Loaded Mile per 100 Lb

Size of truck	Cost per mile per 100 lb
30,000 lb maximum	$0.70

TABLE 8-4
Approximate Installation Cost

Type of member	Cost
Supporting member	30% of cost of fabricated member
Heavy timber decking	$150 per 1,000 fbm

CHAPTER NINE

Moisture Protection

WATERPROOFING

Walls below grade will always require waterproofing. This can be done in several different ways, among which are: 1. using waterproofing paint; 2. plastering; 3. membrane waterproofing; and 4. integral waterproofing.

Waterproof Paint

Waterproofing may be accomplished by painting the walls with one or more coats of water-repellent paint.

One gallon of paint will cover from 60 to 140 sq ft of wall surface depending on the smoothness of the surface and will cost from $2 to $5 per gallon depending on the quality of the paint specified.

A painter should be able to apply from 600 to about 900 sq ft per day. The amount of coverage possible per day will be determined by the working conditions and the roughness and texture of the surface which is to be covered.

Plastering

Waterproofing is often done by applying a coat of waterproof plaster, which may vary from 1/2 to 3/4 in. in thickness.

The material may be a rich portland-cement mortar used alone or mixed with a water resisting compound. The cost of this material will vary from $3 to $4 per 100 sq ft.

The labor required to apply this material will be in the range of from 1 to 3 hr depending on the thickness of the coat required and the condition of the surface.

Membrane Waterproofing

Membrane waterproofing is accomplished by mopping the wall with tar or asphalt, placing a layer of roofing felt against the mopped surface, and then applying another mopping over the felt. One or more layers of felt may be used, with as many as five being employed in extreme cases. There will always be one more coat of tar or asphalt than of felt. Allow from 10 to 15% for lapping of the felt.

From 25 to 40 lb of tar or asphalt are required per 100 sq ft of wall per coat. The cost of this material will vary from $3 to $5 per 100 lb. The felt used will weigh from 15 to 45 lb per sq ft and cost from $3 to $6.

The labor required will vary from 0.7 to 1.3 hr per 100 sq ft per coat for applying the felt and the same per coat for the mopping with tar or asphalt.

Integral Waterproofing

This type of waterproofing is achieved by placing additives in the concrete mix. The cost of such additives will vary over a wide range, say from $2 to $7 or even $8 per cu yd, depending on the product used.

The equipment required for waterproofing will vary due to the procedure used. The equipment could include blowtorches for drying basement walls to receive the treatment, means for liquifying the tar or asphalt, brushes for cleaning the wall, sprayers, small tools, etc. An allowance of $50 to $100 will cover most jobs except for very large work.

Local prices for waterproofing materials vary widely and should be carefully checked for every job.

ROOFING

Roofing prices are usually quoted per square, which designates an area of 100 sq ft. Flashing is figured by the square or lineal foot, the price depending on the material used, such as copper, aluminum, or tin. Gutters and downspouts are priced by the foot.

The most common types of roofing are shingles and built-up roofing.

Wood Shingles

Roofing felt varying in weight from 15 to 30 lb per square is required on the sheathing under the shingles. The cost will range from $3 to $5 per 100 lb and the labor time to place from 0.5 to 1 hr per square. Allow $0.35 per square for roofing nails.

The number of shingles required per square will vary according to the amount of shingle laid exposed to the weather. Table 9-1 will show some approximate figures for differing amounts of exposed shingle.

TABLE 9-1
Number of Wood Shingles Required per Square

Length laid to weather	No. reqd. per square (including waste)	Lb of 3d nails required
4 in.	1,000	3.3
5 in.	800	2.1
6 in.	670	1.6

The cost of wood shingles will vary considerably. They usually come in three grades. Red cedar shingles may vary in cost as follows:

No. 1 grade $8.00 per bundle of 250 shingles
No. 2 grade $5.00 per bundle of 250 shingles
No. 3 grade $2.50 per bundle of 250 shingles

Allow 45¢ per pound for galvanized nails. Allow 10 to 20% for waste in shingles according to the amount of the surface interrupted by ridges, valleys, etc.

A carpenter can lay from about 1,200 to 2,000 shingles per 8-hour day. An experienced shingler can lay about 1/3 more.

Asphalt Shingles

Asphalt shingles are laid in much the same manner as are wood shingles except that they are usually made in strips up to 36 in. long and placed with about the same exposed area as wood shingles and require the underlay of felt.

The cost of these shingles varies a great deal due to the weight of the shingles per square; the higher the weight per square the longer the life of the shingle and the higher the cost. This cost can vary from $5 to $15 per square. Add 5% to the material to allow for ridge and valley strips. Allow $2 per square for nails. Allow 5 to 15% for waste.

The labor required to lay a square of asphalt shingles will vary from 2 to 4 hr for carpenters. When experienced shinglers are used for this work the time may be reduced about 20%.

There are a great many more types of shingles, such as: asbestos, metal, slate, tile, etc.; however, it is not in the scope of this text to cover all possible types. Our main purpose is to teach procedure. Asphalt and wood shingles were chosen for discussion since they are the most commonly used.

Built-up Roofing

Built-up roofing consists of from two to five or more layers of roofing felt with alternate layers of tar or asphalt. The felt will vary in weight from 15 to 40 lb per 100 sq ft. From 20 to 30 lb of tar or asphalt are required per coat per square of roof area. When gravel surfacing is used twice the above amount of tar or asphalt is required for mopping over the top layer of felt. Allow 10% of the felt area for waste and lapping.

The gravel surfacing will require from 250 to 450 lb per square as specified and can cost from $5 to $20 per ton. The felt will cost from $3 to $6 per 100 lb. The tar or asphalt will cost from $3 to $5 per 100 lb.

The labor required will be:

For laying felt per 100 sq ft	0.2 to 0.4 hr
For mopping tar or asphalt per 100 sq ft	0.2 to 0.4 hr
For mopping last coat and gravel 100 sq ft	0.5 to 0.8 hr

For equipment allow about $50 for small tools plus cost of conveyor or crane and heating equipment.

FLASHING

Flashing is waterproof material placed around joints in roof and walls to keep out water, such as at ridges, hips and valleys, where roof meets walls, parapets, or chimneys, or around openings in the roof; also over doors and windows.

Flashing is usually measured by the linear foot for widths up to 12 in. and in square feet for widths greater than 12 in.

Flashing is made of many different materials such as galvanized steel, tin, aluminum, copper, and treated fabrics. The costs of several types of flashing are shown on Table 9-2.

DRAIN TILE

Drain tile is often used in conjunction with membrane waterproofing to keep basements dry. The tile is laid around the outside of the wall footing with the invert of the tile level with the bottom of the footing. The tile is laid in a bed of gravel and drained to the storm sewer. Tiling similarly laid is used to drain athletic fields, airports, cultivated fields, etc.

To estimate the cost of this type of waterproofing it is necessary to ascertain the local prices of tile and stone, excavating equipment costs, labor, etc., and then proceed as outlined in the text for the completion of any estimate.

TABLE 9-2
Material Costs and Labor Necessary to Install
Several Types of Flashing

Item	Material cost	Labor to Install
Ridge strips 26 to 28 gal. galv. steel	$20 to $35 per 100 lin ft	2 to 4 hr
Gutters - 4″ to 6″ Downspouts — 3″ to 4″ 26 to 28 gal. galv. steel	$20 to $40 per 100 lin ft	6 to 8 hr
Gutters — 5″ aluminum Downspouts—aluminum	$25 to $35 per 100 lin ft	6 to 8 hr
Gutters — 4″ to 6″ copper	$125 to $135 per 100 lin ft	6 to 8 hr
Copper flashing 16 to 20 oz	$1.00 to $1.50 psf	0.25 to 0.35 hr
Aluminum flashing 0.032 to 0.05 in. thick	$0.30 to $0.40 psf	0.2 to 0.3 hr
Gravel stops 0.02 in. thick aluminum with 4-in. high face	$30 to $35 per 100 lin ft	6 to 7 hr
Gravel stops 16 oz. copper with 3 in.-to 6 in.-high face	$120 to $150 per 100 lin ft	8 to 9 hr
Reglets for counter flashing 10 to 16 oz copper	$30 to $90 per 100 lin ft	4 to 6 hr

CHAPTER TEN

Glazing

ENTRANCES & GLASS WALLS

Glass can be roughly divided into window or sheet glass and plate glass. The costs per square foot of a few types of glass are shown in Table 10-1.

There are many different types of thermopane glass made by assembling several layers of glass into one unit. This glass is very expensive and varies in cost per sq ft from about $3.00 to $15.00 or more.

Aluminum trim is usually used in making glass doors and show windows or large wall sections of glass. This tubular trim comes in sizes such as 1 3/4″ × 4″, 1 3/4″ × 4 1/2″, etc., and is priced at about $2.00 per lin ft.

TABLE 10-1
Approximate Cost of Glass per Sq Ft

Thickness		Cost per sq ft	Glazing compound per sq ft
	SHEET GLASS		
1/8″	double strength (window glass)	$0.65	(0.5 lb) $0.15
1/8″	obscure	0.75	(0.5 lb) 0.15
7/32″	obscure	0.95	(0.5 lb) 0.15
1/4″	obscure	1.05	(0.5 lb) 0.15
	PLATE GLASS		
1/4″	plate glass — clear	1.00	(0.2 lb) $0.06
1/4″	plate glass — tinted-heat resist	1.40	(0.2 lb) 0.06
3/8″	plate glass — clear	2.00	(0.2 lb) 0.06
3/8″	plate glass — tinted-heat resist	2.50	(0.2 lb) 0.06
1/2″	plate glass — clear	3.25	(0.2 lb) 0.06
1/2″	plate glass — tinted-heat resist	4.00	(0.2 lb) 0.06

To set plate glass in large sizes will require from 2 to 10 men depending on the type and size of the glass to be handled, and these men including a truck with its driver will have to be in attendance until the glass is set and secured.

Glass sizes are designated by the term "united inch," which is simply the height plus the width of the piece. Table 10-2 shows the number of men who will be required to handle certain sizes of glass.

TABLE 10-2
Labor Required to Handle Glass

Size (in united inches)	No. of men required to handle
140	2
160	3
180	4
200	5
240	6
270	7

When estimating the cost of plate glass installed it is best to get a price from a local firm that specializes in this work. However, we will discuss these prices in general.

The cost of one plate glass door 3' × 7' of 3/8" glass is:

3/8" plate glass—21 sq ft @ $2.00	$ 42.00
Glazing compound—(7' × 3') × 0.2 lb/sq ft × $0.30 per lb	1.26
1 3/4" × 4" aluminum trim—2(7'+ 3') × $2.00 per lin ft	40.00
Hardware	15.00
Material	$ 98.26
To set and glaze, 2 men—5 hr @ $7.35	$ 73.50
Truck and driver, 5 hours @ $10.00	50.00
Labor	$123.50

Each glass front presents a different problem. The cost per square foot of the glass in place will depend upon the number of men necessary to handle the maximum size of glass used, and the type of glass. Table 10-3 shows the labor costs per sq ft for 5 different glass fronts and demonstrates the method used to obtain these prices.

Table 10-3 illustrates the effect that fixed costs have on the labor for setting and glazing plate glass fronts.

TABLE 10-3
Some Labor Costs for Glass Fronts per Sq Ft

Wall size in ft	Sq ft of glass	Hr to set and glaze	Travel time (gang) in hr	Travel time (driver) in hr	Wait time (driver) in hr	Set and glaze @ $7.35	Driver @ $6.20	Truck @ $5.00	Cost per sq ft to set and glaze
38 × 11	418	6,7*	6†	1	7	$352.40	$49.60	$40.00	$ 1.05
44 × 8	352	6, 7	6	1	7	352.40	49.60	40.00	1.25
30 × 9.5	285	6, 7	6	1	7	352.40	49.60	40.00	1.55
30 × 9	270	7, 7	7	1	7	411.00	49.60	40.00	1.85
24 × 9	216	6, 7	6	1	7	352.40	49.60	40.00	2.05

*First figure shows number of men required to handle the glass and the last figure the number of hours for each man.
†Traveling time to job and back — 1 hour for men and truck.

A considerable number of men is required to handle large pieces of glass and a truck with driver has to be on hand to move the men and glass to and from the job. Once on the job the men can set the glass quickly and so place a number of pieces in about the same overall time as one piece. This is the reason that the cost of labor per sq ft for a glass front in place tends to be smaller for the larger jobs.

It must be remembered that the linear feet of trim including mullions should be added to the cost of the glass and glazing compound, in computing the material cost of plate glass. The time shown in the table for labor includes an allowance for placing the trim.

DOORS

There are many types of steel and fire doors in a wide range of prices which will have to be determined locally for each type of door used.

Table 10- 4 lists the price of several types of wood doors.

TABLE 10-4
Approximate Cost of Wood Doors

| | Flush type | | Panel type | |
| | Birch | Pine | Pine | Pine |
Door size	Hollow core	Solid core	Plain	Glazed
2/0 × 6/8 × 1 3/8	$ 13		$ 23	
2/4 × 6/8 × 1 3/8	14		24	
2/6 × 6/8 × 1 3/8	15		25	
2/8 × 6/8 × 1 3/8	16		28	
3/0 × 6/8 × 1 3/8	17		31	
2/8 × 6/8 × 1 3/4	18	$ 65	37	$ 59
3/0 × 6/8 × 1 3/4	20	70	41	62
3/0 × 7/0 × 1 3/4	22	75	45	65

Notes: Add $60 to above price for prehung exterior and $50 for prehung interior doors, which will include frame trim and hardware. Add $15 for weatherstripping — bronze type. Add $25 for screen.
For labor to hang doors, see Tables 7-4 and 7-5.

WINDOWS

Metal sash are, in most cases (especially in industrial work), glazed after the sash has been installed. The following example shows the approximate cost of glazing, per square foot of glass, of stock sizes of metal sash:

Double-strength glass (1/8″)	$0.65
Glazing compound 0.5 lb @ $0.30	0.15
Material	$0.80
Labor to glaze	$0.50

Tables 10-5 and 10-6 show approximate costs of unglazed steel and aluminum sash of several different types.

Table 10-7 shows the approximate cost of several different types and sizes of prehung wood windows.

SKYLIGHTS

Skylights will not be discussed in detail in this text. They are very expensive and their local cost should be obtained for each job.

The cost of skylights per square foot of clear opening will vary from about $3.00 to $5.00 for materials, and $1.20 to $2.00 for labor to install.

TABLE 10-5
Approximate Cost of Unglazed Steel Sash per Sq Ft

Type	Material	Labor to install
Industrial projected (stock)	$1.50 to $3.00	$1.00
Industrial horizontally pivoted (stock)	1.30 to 2.50	1.00
Industrial fixed (stock)	0.90 to 1.50	1.00
Steel sash (custom made)	5.00 to 8.00	1.00 to 2.25
Mullions (as required)	1.00 to 1.50 per lin ft	1.00 per lin ft

TABLE 10-6
Approximate Cost of Unglazed Aluminum Sash per Sq Ft

Type	Material	Labor to install
Projected (stock)	$4.00 to $5.00	$1.00
Fixed (stock)	1.75 to 2.75	1.00
Double hung (stock)	2.00 to 3.00	1.00
Aluminum sash (custom made)	8.00 to 12.00	1.00 to 2.25
Mullions (as required)	1.50 to 3.00 per lin ft	1.00 per lin ft

TABLE 10-7
Approximate Cost of Prehung Wood Windows
(Including screens, frames, and trim)

Size and type		Material cost
2' - 8" × 4' - 6"	Doublehung wood	$40.00*
3' - 0" × 4' - 6"	Doublehung wood	42.00*
3' - 4" × 4' - 6"	Doublehung wood	46.00*
1' - 10" × 3' - 2"	Wood casement — 1 Leaf	65.00
3' - 10" × 4' - 2"	Wood casement — 2 Leaf	135.00

*Add $10.00 for weatherstripping with bronze.
Add $50.00 if metal storm windows are required.

Notes: Average carpenter labor per hour - $7.14.
See Table 7-5 for labor to install.

CHAPTER ELEVEN

Finishes

LATHING AND PLASTERING

Wood lath is seldom if ever used any more and for this reason will not be considered in our discussion. The laths most commonly used are metal lath and gypsum or fibre or plasterboard lath. We will use only metal and gypsum lath in our examples. The measurement generally used for lathing and plastering is the square yard. If the lathing and plaster for a 12′ × 20′ × 8′ room is to be estimated, the following areas are used:

$$
\begin{array}{ll}
\text{For walls } (12' + 20')\ 2' \times 8' & = 512 \text{ sq ft} \\
\text{For ceiling } (12' \times 20') & = \underline{240 \text{ sq ft}} \\
752 \div 9 = 83.6 \text{ sq yd total} & 752 \text{ sq ft}
\end{array}
$$

Do not deduct anything for openings except in unusual cases, such as store fronts where twice the actual area of plaster bordering the openings should be used; or when one wall is not plastered and should of course not be included. Where beams project below the ceiling line allow a foot-wide strip the length of the beam extra for each internal or external angle. Making no deductions for openings will usually compensate for corner beads, moldings, etc., which we will not consider here.

Tables 11-1 through 11-6 show some approximate costs of lath and plaster.

TABLE 11-1
Approximate Cost of Gypsum Lath per 100 Sq Yd

Item	Material	Labor
3/8″ lath nailed to wood studs	$45.00	9.5 hr
10 lb nails @ $0.30	3.00	
Total	$48.00	9.5 hr

TABLE 11-2
Approximate Cost of Expanded Metal Lath
per 100 Sq Yd

Item	Material	Labor
3.4 lb metal lath nailed to wood studs	$55.00	11.5 hr
Fasteners	5.00	
Total	$60.00	11.5 hr

Note: If metal studs are used in Table 11-2, fasteners will
cost $2.00 more and labor will be 15 hr instead of 11.5.

TABLE 11-3
Approximate Cost of Suspension System for Metal
Lath per 100 Sq Yd

Material	Labor
$120	29.5 hr

Note: Labor time in Tables 11-1, 11-2, and 11-3 is for
lather @ $7.80 per hour.

TABLE 11-4
Approximate Cost of Gypsum Plaster per 100 Sq Yd

Grounds	No. of coats	Material	Labor
5/8"	2	$65.00	25 hr
3/4"	3	75.00	30 hr

TABLE 11-5
Approximate Cost of Vermiculite or Perlite
Plaster per 100 Sq Yd

Grounds	No. of coats	Material	Labor
5/8"	2	$60.00	24 hr
3/4"	3	90.00	30 hr

Note: Labor time in Tables 11-4 and 11-5 includes plasterer
with helper = $7.75 + $6.50 = $14.25 per hr.

TABLE 11-6
Approximate Cost of Stucco per 100 Sq Yd

Total thickness	No. of coats	Material	Labor
1″ on wood lath*	3	$160.00	30 hr
1″ on masonry	3	60.00	8 hr

Note: Labor time includes plasterer with helper and lather =
$7.75 + $6.50 + $7.80 = $22.05.
*Material includes wood lath.

DRYWALL CONSTRUCTION

In drywall construction the walls are covered with gypsum wallboard, wood paneling, drywall tile, etc., and the ceilings with gypsum wallboard or with some form of tile (the acoustical tiles being most popular).

In taking off material for drywall construction use the same procedure as for plaster. By not deducting the wall openings you will allow for material waste, fitting around openings, cost of nails, fastenings, etc.

Tables 11-7 through 11-11 show the costs of several types of drywall construction.

TABLE 11-7
Approximate Cost of Gypsum Drywall per 100 Sq Ft

Type	Location	Material	Tape	Labor
3/8″	Walls	$7.00	$1.00	1.8 hr
1/2	Walls	9.00	1.00	2.0 hr
5/8	Walls	10.00	1.00	2.3 hr
3/8	Ceiling	7.00	1.00	2.1 hr
1/2	Ceiling	9.00	1.00	2.3 hr
5/8	Ceiling	10.00	1.00	2.5 hr

TABLE 11-8
Approximate Cost of Paneling per 100 Sq Ft

Item	Material	Labor
1/4″ Birch (4′ × 8′ sheets)	$ 25 to $ 75	5 hr
1/4″ African mahog. (4′ × 8′ sheets)	50 to 65	5 hr
1/4″ Lauan mahog. (4′ × 8′ sheets)	10 to 20	5 hr
1/4″ Oak, walnut, or cherry (4′ × 8′ sheets)	30 to 85	5 hr
1/4″ Rosewood or chestnut (4′ × 8′ sheets)	175 to 225	5.5 hr
1/4″ Teak (4′ × 8′ sheets)	80 to 100	5.5 hr
3/4″ Knotty pine	25 to 35	6.3 hr
3/4″ Aromatic cedar closet lining	25 to 38	6.3 hr

TABLE 11-9
Approximate Cost of Acoustical Tile per 100 Sq Ft

Item	Material	Labor
Tile	$25 to $50	$20.00
Ceiling prime	3.00	
Ceiling cement	6.00	
Total	$34 to $59	$20.00

TABLE 11-10
Approximate Cost of Suspension for
Drywall Ceiling per 100 Sq Ft

Material	Labor
$10 to $14	1.5 to 2.5 hr

TABLE 11-11
Approximate Cost of Furring per 100 Sq Ft

Location	Material	Labor
Wall	$4.00	0.5 hr
Ceiling	4.00	1.5 hr

Note: Labor time in Tables 11-7 through 11-11 are for average wage of carpenter gang of $7.14 per hour.

RESILIENT FLOOR TILE

Resilient floor tile includes asphalt tile, rubber, cork, linoleum, vinyl asbestos, vinyl tile, and vinyl sheet flooring.

Each type of resilient tile has its own special uses and each is made in several different grades. Prices vary a great deal. Table 11-12 is shown as an example and should not be used for actual work. Local prices should always be ascertained.

In taking off area to be covered use gross areas and allow for waste:

15% for areas under 50 sq ft
10% for areas 50 to 100 sq ft
8% for areas 100 to 200 sq ft
6% for areas 200 to 300 sq ft
5% for areas 500 to 1000 sq ft
3% for areas 1,000 to 10,000 sq ft
2% for areas over 10,000 sq ft

TABLE 11-12
Approximate Cost of Various Types of Resilient
Floor Tile per 100 Sq Ft (except as noted)

Type	Material cost	Labor hours
Asphalt tile on concrete, 9″ × 9″ × 1/8″	$ 18.00	1.00
Asphalt tile on concrete, 9″ × 9″ × 3/32″	15.00	1.00
Base of rubber (cost per lin ft), 1/8″ thick × 4″ high	00.20	0.02
Base of rubber (cost per lin ft), 1/8″ thick × 6″ high	00.30	0.02
Linoleum, standard weight	35.00	2.50
Linoleum, heavy (1/8″)	45.00	2.50
Rubber, 1/8″ thick	50.00	1.00
Rubber, 1/4″ thick	80.00	1.00
Vinyl, 9″ × 9″ or 12″ × 12″ −0.05″ thick	35.00	1.00
Vinyl, 9″ × 9″ or 12″ × 12″ − 0.08″ thick	100.00	1.00
Vinyl, 9″ × 9″ or 12″ × 12″ − 1/8″ thick	120.00	1.00
Vinyl base (cost per lin ft) 4″ × 1/8″	000.14	0.02
Vinyl base (cost per lin ft) 6″ × 1/8″	000.20	0.02

Notes: These prices are average for the different types.
They vary due to thickness, color, and grade. Local prices
should be obtained for each job.
The labor time is for tile layer with helper.

Tile layer	$ 7.55
Helper	6.30
	$13.85 per hr

Ceramic tile provides a hardwearing, long lasting surface which is almost maintenance free. Many colors, sizes, shapes, and finishes are available. Some of the most commonly used ceramic tiles are: glazed ceramic wall tile, ceramic mosaic tile, and quarry tile.

There are many variables which determine the cost. Ceramic wall tile, for instance, can be unmounted or back mounted, and installed by the "mud-set" portland cement mortar method, or by the use of various types of adhesives, or by the comparatively new "dry-set" portland cement mortar method (the last method being perhaps the cheapest).

Ceramic tile is estimated by the square foot and the trim, such as base, cap, etc., by the linear foot. All window and door openings should be deducted and the trim pieces necessary to finish the openings added.

Plastic wall tile is sometimes used because of economy. It will usually cost from 1/2 to 2/3 the price of ceramic tile installed but does not have the lasting qualities of the ceramic article.

TERRAZZO

Terrazzo is a poured-in-place surface over a concrete base. On this surface is poured an underbed of sharp screened sand in portland cement and covered with a topping of 1/2 in. to 3/4 in. of granulated marble of mixed selected colors. Metal strips are usually used to break up the surface in large squares. Preassembled decorative units are often embedded in the surface. The surface is then ground to a fine finish.

There are several methods of constructing a terrazzo floor and the cost varies considerably due to the type of floor and the method of placing. One price is shown in Table 11-13.

TABLES 11-13
Approximate Cost of Ceramic Tile Floors and Walls
per 100 Sq Ft (except as noted)

CERAMIC TILE (Figures include bed)

Location			Material cost	Labor hours
	Floors — porcelain tile	1″ × 1″	$74.00	06.90
(Mud-set)	Walls — glazed	4 1/4″ × 4 1/4″	64.00	11.00
(Mud-set)	Walls — glazed	6″ × 4 1/4″	70.00	10.00
(Mud-set)	Cove base (cost per lin ft)	4 1/4″ × 4 1/4″	00.62	00.11
(Mud-set)	Cove base (cost per lin ft)	6″ × 4 1/4″	00.52	00.10

QUARRY TILE (Figures include bed)

Location			Material cost	Labor hours
(Mud-set)	Floors —	4″ × 4″ × 1/2″	$76.00	11.00
(Mud-set)	Floors —	6″ × 6″ × 1/2″	63.00	8.40
(Mud-set)	Floors —	9″ × 9″ × 3/4″	80.00	00.10
(Mud-set)	Cove base (cost per lin ft)	2″ or 5″ high	00.48	00.20

TERRAZZO (Figures include bed, dividing strips, etc.)

	Material cost	Labor hours
Floors — 5/8″ topping bonded to 1 1/8″ conc. bed	$48.00	9.00

Note: For small jobs, such as houses, the labor costs should
be doubled.

The labor time is for tile setter with helper:

Tile setter	$ 7.70
Helper	6.30
	$14.00 per hr

Plastic wall tile in the 4 1/4″ × 4 1/4″ to 9″ × 9″ sizes is
available at about 2/3 the cost of ceramic tile.

PAINTING

Painting is a rather complex problem. There are some surfaces, such as at
walls and cornices, which will require the same amount of paint per square
foot of surface but where the labor will be much higher for the area covered
on the cornice because it is much harder to reach.

Some surfaces will require more paint to cover than others. Some areas will require a great deal of putty for patching holes and imperfections in the wood. New surfaces will require different preparation for painting than areas where the paint is being renewed. Some jobs will require a great deal of preparation such as removing paint which is badly checked or scaled.

The cost of painting will vary a great deal with the cost of the paint used and the number of coats specified.

In taking off areas for painting do not deduct for doors and windows except for large plate glass doors and windows. Then allow for trim as follows:

> Allow 2 sq ft for every running foot of trim
> Allow 3 times area of window out to out of trim
> Allow 2 times area of door opening out to out of trim

Tables 11-14, 11-15, and 11-16 are self-explanatory and are based on the use of paint brushes. However, paint is now, in some instances, being applied with a roller. More paint per coat is required when it is applied by roller. However, in some cases the painting can be accomplished more quickly by employing the roller, such as:

Paint on wood siding	—	Approx. 25% faster
Stain on wood siding	—	Approx. 25% faster
Paint on dry wall or plaster	—	Approx. 50% faster
Paint on masonry	—	Approx. 50% faster

TABLE 11-14
Some Average Costs of Paint

Paint	Avg cost per gal	Paint	Avg cost per gal
Interior varnish	$9.40	Exterior spar varnish	$7.30
Interior latex or oil base, flat	5.40	Waterproof, silicone	6.80
Enamel	8.90	Turpentine	3.10
Exterior — oil base	7.50	Shellac	3.40
Exterior — oil base (1 coater)	9.50	Aluminum	3.80
Shingle stain	5.00	Asphalt	1.00
Shingle paint	8.00	Asphalt, roof, & gutter	1.15
Masonry paint, interior	7.20	Creosote	1.25
Primer—plaster, wood, metal	4.60		
Primer — for masonry	5.40		

Note: Above prices are figured on minimum orders of 25 gal.
Lots of 100 gal will cost about 10% less per gal.

It should be remembered that to the above must be allowed time for trimming out the corners and close places with a brush.

Some equipment cost should be allowed for such items as brushes, putty, ladders, and scaffolding where required. From $25 to $50 should be sufficient for a small job where little scaffolding is required. This price will rise sharply for larger jobs requiring an appreciable amount of scaffolding.

TABLE 11-15
Average Paint Coverages and Time of Application

Location	Coats	Area per gal in sq ft using brush	Area per man-hour in sq ft using brush
1.* Varnish — hardwood floor	One coat	400	140
2. Varnish on trim	Sealer	300	125
	Next coat	330	150
3. & 5. Paint on dry wall or plaster	Prime	440	200
	Next coat	450	170
6. Paint on interior woodwork	Prime	325	180
	Next coat	375	160
7. Paint on interior wood trim	Prime	325	100
	Next coat	375	135
8. Paint on doors and windows	Prime	350	140
	Next coat	400	150
9. Paint on wood siding	Prime	250	120
	Next coat	300	160
10. Stain on shingle siding	First coat	325	130
	Next coat	400	160
11. Paint on shingle siding	Prime	300	120
	Next coat	375	150
12. Paint on exterior wood trim	Prime	300	80
	Next coat	350	90
13. Paint on masonry	Prime	200	100
	Next coat	225	150

*Cross reference numbers to Table 11-16.

TABLE 11-16
Some Approximate Painting Costs per Sq Ft

Location	Type	Coats by brush	Units	Material	Labor
		INTERIOR WORK			
1.* Hardwood floor	Varnish	1 Coat (no sanding)	sq ft	$0.02	$0.05
2. Wood trim – interior	”	1 Coat & sealer	sq ft	0.04	0.11
3. Dry wall or plaster	Latex	Primer & 1 coat	sq ft	0.02	0.08
”	”	” & 2 ”	sq ft	0.04	0.12
5. Dry wall or plaster	Enamel	Primer & 1 coat	sq ft	0.03	0.08
”	”	” & 2 ”	sq ft	0.05	0.13
6. Woodwork	Enamel	Primer & 1 coat	sq ft	0.04	0.09
”	”	” & 2 ”	sq ft	0.06	0.13
7. Wood trim – interior	Enamel	Primer & 1 coat	sq ft	0.04	0.13
”	”	” & 2 ”	sq ft	0.06	0.18
8. Doors & windows	Enamel	Primer & 1 coat	sq ft	0.04	0.10
”	”	” & 2 ”	sq ft	0.06	0.15
13. Masonry	Acrylic	Filler & 1 coat	sq ft	0.06	0.12
”	”	” & 2 ”	sq ft	0.09	0.17
		EXTERIOR WORK			
Blinds	Oil base	Primer & 1 coat	each	$1.00	$4.25
Screens	”	” & 1 ”	each	0.50	2.75
9. Wood siding	Oil base	Primer & 1 coat	sq ft	0.04	0.11
”	”	” & 2 ”	sq ft	0.07	0.15
10. Wood shingles	Stain	Sealer & 1 coat	sq ft	0.03	0.10
”	”	” & 2 ”	sq ft	0.05	0.15
11. Wood shingles	Paint	Primer & 1 coat	sq ft	0.04	0.11
”	”	” & 2 ”	sq ft	0.06	0.16
12. Wood trim – exterior	Oil base	Primer & 1 coat	sq ft	0.04	0.17
”	”	” & 2 ”	sq ft	0.06	0.25

Note: Time shown above for labor includes allowance for puttying.
*Cross reference numbers to Table 11-15.

CHAPTER TWELVE

Prefabricated Materials

CURTAIN WALL CONSTRUCTION

Cast Panels

The cast panel is a relatively new type of curtain wall material that is being extensively used for all types of buildings. The panels are made of concrete and can be cast in a wide variety of sizes and shapes ranging from pieces one story in height to two or more stories cast as solid panels or to include framed openings for windows.

Because of the high cost of the forms in which these panels are cast there must be many pieces made from each pattern if this type of construction is to be economical. Jobs containing a minimum of 5,000 to 6,000 sq ft will usually be economical. Jobs half this size will probably cost twice as much per sq ft.

Table 12-1 shows the approximate costs per sq ft for plain panels without window openings.

TABLE 12-1
Approximate Cost of Solid Cast Panels per Sq Ft

Thickness of panel (in.)	Size range (sq ft)	Base cost (sq ft)	Add to base cost Irreg. surface	Exposed agg.	Special finishes
3	48	$3.00	$1.00	$0.40 - $2.00	$0.75 - $2.25
4	144	3.50	1.00	”	”
5	200	3.75	1.00	”	”
6	310	4.25	1.00	”	”
7	310	4.50	1.00	”	”
8	310	5.00	1.00	”	”

The base cost per sq ft for window panel units and other complicated forms will vary from approximately $4 to $12 or more. There is a considerable cost involved in delivering these panels from the casting yard to the job site. When this distance is not over 50 miles the cost will range from $0.30 to $0.75 per sq ft.

The erection, plumbing, and aligning of these panels will require 4 or 5 men and a crane. Such a crew should erect from 25 to 150 sq ft per hour depending on the size and shape of the panels, the job conditions, and the location in the building. With a crew of 4 carpenters and a foreman the cost per hour would be:

4 carpenters @ $7.85	=	$31.40
1 foreman @ 8.30	=	8.30
Labor	=	$39.70
Crane @ $35/hr	=	35.00
Total		$74.70

The cost per square foot will vary from $0.50 to $3.00. This price will include all direct labor and equipment costs for the erection, plumbing, and alignment of the panels to which must be added overhead and profit.

There still remains the cost of caulking between all panels, which usually is required on both faces. This can be priced at about $0.25 material and $0.50 labor per lin ft for each face of panel.

The two principal advantages of this type of curtain wall are the wide variety of designs possible and the speed of erection. For instance, a 4' x 8' cast stone panel can be bolted into place much faster than the same area of stone can be placed by stone setters. The time saved can be very important in the construction of a large building.

These prices are only approximate. There are many factors which control the price per sq ft of cast panels for every job is different because of the amount of repetition of typical panels and the complexity of the design. It is therefore imperative to obtain a figure from the supplier before any firm bid price can be made.

To illustrate the difficulty in pricing this material let us look at the cost items listed on one of the bid sheets for 55,000 sq ft of paneling for a church. Table 12-2 is a typical pricing sheet for panels made of concrete.

Then factory overhead, hauling, and profit were added. The final cost of the type of panel in this particular job would be from $5 to $7 per sq ft in different areas of the United States, depending for the most part on wage differentials and practices.

TABLE 12-2
Typical Pricing Sheet for Concrete Panels

Material	*Labor*	
1. Facing	1. Drafting	14. Inspection
2. Backing	2. Patterns	15. Form dressing
3. Mesh	3. Form setting	16. Placing tile
4. Reinforcing	4. Cut mesh	17. Grouting tile
5. "A" Anchors	5. Mixing	18. Miscellaneous
6. Pls. 4″ × 1/4″ × 4′	6. Casting	
7. Wire loops	7. Troweling	
8. 3/4″ wedges	8. Take out	
9. Cable loops	9. Finishing	
10. Coil loops	10. Handling	
11. P.V.C. reglets	11. Siliconing	
12. Form lumber	12. Loading	
13. Steel mold	13. Hauling	

The 31 items of expense do not make any allowance for the varying size of the panels and the special engineering, where necessary, to comply with special architectural requirements. This further emphasizes the need to secure bids from panel manufacturers before attempting to estimate such work.

Metal Panels

Exterior wall panels are sandwich style with the outer skin of aluminum or enamel-covered metal enclosing from 1 in. to 2 in. of insulation of fiberglass, foam glass, polystyrene, polyurethane, etc.

The material cost of these sandwich panels in lots of over 5,000 sq ft will vary from $3 to $10 or more per sq ft. The price variation is due to kind and thickness of the outer skin, and the type and thickness of the insulation. In smaller lots the price will increase considerably. To this price must be added the cost of supporting framework and 2 to 5% for flashing. The erection cost will run about $0.50 to $1.50 per sq ft. Because of the many variables the price of sandwich panels should be ascertained from the manufacturer for each particular job.

The same advantages that apply to cast panels made of concrete will accrue to these panels.

PRECAST FRAMING

Precast concrete framing members are playing an increasingly important role in construction. Any material that can be precast in quantity in the shop instead of in the field enjoys a distinct price advantage. Any work that can be done in the shop always costs less than the same work performed in the field. These members are usually priced per sq ft and come much cheaper in large amounts, say in excess of 10,000 sq ft.

Double and Single Tees

Prestressed tee sections (see Figure 12-1) are used for both roof and floor framing to carry long spans without intermediate supports, thus giving large column-free areas. These tees are in effect precast beam and slab sections and are much cheaper than cast-in-place framing.

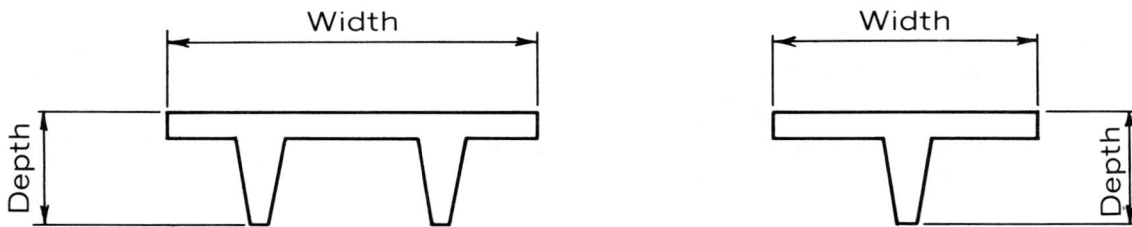

DOUBLE TEE   SINGLE TEE

FIGURE 12-1
Typical Tees Used for Roof and Floor Framing

Table 12-3 shows some sample costs of tee sections for quantities of 10,000 sq ft or more.

The delivery of the tees from the plant to the job will cost about $0.10 per sq ft per 50-mile haul.

The erection will probably cost:

For double tees $0.30 to $0.60 per sq ft

For single tees $0.40 to $0.90 per sq ft

TABLE 12-3
Some Approximate Costs of Tees per Sq Ft

Type section	Place used	Depth (in.)	Width (ft)	Span (ft)	L.L. per sq ft	Cost per sq ft
Double T	Roof	12 to 36	4 to 8	30 to 90	30	$1.50 to $2.50
Double T	Floor	12 to 36	4 to 8	20 to 60	100	1.50 to 2.50
Single T	Roof	26 to 46	4 to 8	40 to 100	30	2.00 to 4.00

There is also the cost of welding the connections that tie the tee flanges together, which will probably cost $3 to $5 per tee. The connections at the ends of tees over beams or columns will have to be grouted at a cost of $10 to $20 each.

Beams and Columns

Tee sections are usually carried by simply supported concrete beams, which in turn are carried by concrete columns, all of which are frequently precast.

The supporting beams when prestressed as well as precast will cost between $140 and $225 per cu yd. The cost of delivery will be about $6 per ton per 60-mile haul. Precast, prestressed columns will cost $175 to $325 per cu yd.

Erection will cost from $30 to $70 per cu yd including connections.

LIFT SLAB CONSTRUCTION

Lift slab construction is used in buildings up to 16 stories where conditions make it economical. This procedure is patented and the royalties make it so expensive that it is not feasible to use in many instances.

There are some very distinct advantages that accrue to this system of construction. The columns are first erected and the slab on grade poured. Then the upper-floor slabs are poured directly on top of the first slab. They are separated from one another by a membrane to prevent them from bonding to each other. Forms around each column stop the floor so that it does not touch the column. The slabs are then lifted one at a time by jacks placed on the top of each column.

Before each floor is lifted almost all of the building material needed on that floor is placed on the slab and the whole lot pulled up into place. This saves the time and expense of lifting this material by crane or elevator. When the building site is located in a heavy traffic area this system is ideal. The materials required by the several trades can be moved in at night, when the traffic is light, and loaded on the slabs to be lifted next day—thus not interfering with the traffic in the busy daylight hours.

The cost of such slabs will vary between $1.20 and $3.00 per sq ft. This figure includes material, forming, placing, reinforcing, and pouring the concrete. The cost of lifting and securing the slabs in place will run from $0.50 to $1.00 per sq ft including royalties.

TILT-UP CONCRETE PANELS

The chief advantage of tilt-up slabs is the low cost of building the forms and placing the reinforcing bars. The forms are built on the floor, a much faster and cheaper process than fabricating them upright. The floor forms the bottom of the panel thus saving form material. Also the costs of the customary form ties are saved.

Walls which are 5 1/2-in. thick will vary in cost from about $1.50 to $2.00 per sq ft. The erection cost will average $0.25 per sq ft. The connections between panels will cost $2.00 to $5.00 per lin ft.

CHAPTER THIRTEEN

Special Equipment

BLACKBOARDS AND BULLETIN BOARDS

Table 13-1 shows approximate costs of certain types of blackboards. Bulletin boards will cost in the range of $0.40 to $1.50 per sq ft for material and require 0.05 hr of carpenter labor per sq ft for installation.

TABLE 13-1
Some Approximate Costs of Blackboards in Place

Type	Size	Unit	Material cost	Labor hr
Cement asbestos	3/16″ thick	sq ft	$1.25 to $1.50	0.07
Hardboard	1/4″ ″	sq ft	0.59 to 0.70	0.07
Hardboard	1/2″ ″	sq ft	1.00 to 1.10	0.07
Metal on 1/4″ plywood	24 Gauge (Ga.)	sq ft	2.00 to 2.50	0.07
Slate up to 4′ – 6″ wide	3/8″	sq ft	2.00 to 2.50	0.08
Chalk trays	Aluminum	lin ft	1.50	0.08
Trim	″	lin ft	0.50	0.06

Note: Labor refers to the carpenter hours required to install the boards.

CHUTES

The cost of chutes will vary due to many factors. Some costs are as shown on Table 13-2.

TABLE 13-2
Approximate Cost of Chutes per Floor Installed

Type	Material	Labor hr
Laundry — 24"-dia. aluminum	$200	0.8
Laundry — 24"-dia. stainless steel	300	0.8
Mail chutes — aluminum	240 - 400	0.8
Mail chutes — bronze	340 - 500	0.8

Note: Labor refers to the metal worker hours required to install the chutes.

VAULTS

Vaults are usually constructed of concrete walls and slabs except for the entrance door. The walls and slabs are from 6" to 12" thick with steel reinforcing placed at center of the slab spaced about 6 in. centers both ways. The bars are usually 1/2", 5/8", or 3/4" in diameter.

The cost of the vault door and frame will vary with the hours of fire rating required and the size of the opening (see Table 13-3).

TABLE 13-3
Approximate Cost of Vault Doors Installed

Size	Fire rating in hours	Material	Labor hr
2'-8" X 6'-6"	1	$ 500	12
2'-8" X 6'-6"	4	950	18
2'-8" X 6'-6"	6	1100	22
3'-4" X 4'-6"	4	1200	150

Note: Labor refers to the steel worker hours required for the job.

REVOLVING DOORS

Revolving doors are very expensive because of the complexity of the work. The cost will vary from job to job. Such items are generally furnished installed by a subcontractor. The price then would contain the subcontractor's overhead and profit and would thus cover the payroll taxes and insurance for this item. The general contractor would only need to add a small handling charge and some profit.

The cost of revolving doors per opening could vary from $4,000 to $7,000 or more.

KITCHEN EQUIPMENT

Kitchen equipment, like revolving doors, will usually be installed by a subcontractor. These items are specialized and usually custom-made for each particular job. It is therefore necessary to obtain prices for each different piece of work; however, Table 13-4 shows some of this information.

The items of special equipment discussed here cover only a very small part of this field, but will give some idea of the complexity of the problem encountered in making a complete estimate.

TABLE 13-4
Approximate Cost of Some Items of Kitchen
Equipment Installed

Item	Price installed
Coffee urns — 30 gal	$1,200 to $1,800
Coffee urns — 100 gal	1,900 to 3,000
Broilers	800 to 1,600
Coolers — 8′ long	700 to 1,300
Food warmer	125 to 200
Griddle — 3′ long	400 to 900
Food mixers — 20 to 50 qt	600 to 1,400
Freezers — 40 cu ft	1,300 to 1,800
Fryer	200 to 800
Ice cube maker (100 lb per day)	400 to 600
Range, restaurant type	350 to 800
Refrigerators — 60 cu ft	1,500 to 2,000

CHAPTER FOURTEEN

Walks, Drives, and Parking Lots

WALKS

There are a number of different types of walks that are used around residences, schools, institutions, and industrial buildings. In computing the cost of walks it is necessary to allow for rough and fine grading and curing in addition to the approximate costs shown in Table 14-1.

Perhaps the most economical type of walk is made with bituminous concrete on grade using no base. For this type of construction the subgrade must be stable, which means that it must be composed of nonplastic material which will retain firmness under adverse moisture conditions and be well drained. If the subgrade must be stabilized by bringing in special material which must be mixed in with the existing soil and then rolled, the price quoted in Table 14-1 must be increased to cover this cost. This increase will amount to approximately 100%.

TABLE 14-1
Approximate Cost of Walks per Sq Ft

Type	Material	Labor hr
2″ bituminous conc. on grade — no base	$0.18	0.02
Brick on sand bed — flat	0.45	0.13
Brick on sand bed — on edge	0.60	0.22
4″ concrete with mesh on grade — no base	0.30	0.03
5″ concrete same as above	0.34	0.04
6″ concrete same as above	0.42	0.05
2″ precast concrete	0.60	0.06
1″ flagstone	0.40	0.25

Note: Labor refers to the hours of skilled labor required to install the work.

Walks made of brick laid in a sand bed are very decorative and long lasting if properly laid. When the walk has to be disturbed for subsurface work, it is much simpler and cheaper to replace than concrete. However, this is one of the most expensive types of walks, as shown in Table 14-1. If the walk must be laid on a concrete bed and the joints between brick filled with concrete, then add to the cost listed in Table 14-1 $0.30 per sq ft for material and $0.07 per sq ft for labor.

Of all the types of walks concrete is by far the most often used. It never grows sticky in hot weather as does asphalt and is the cheapest paving of all for walks (with the exception of asphalt).

Precast concrete blocks are sometimes used for the paving of walks. This type of paving is more expensive than cast-in-place concrete but does give a pleasing appearance and has the advantage of not requiring time for curing before it can be used. Flagstones are often selected for paving in order to achieve architectural effects. This walk will cost about the same as brick laid flat if the 1″ thickness is used. For thickness of material over 1″ the price will rise sharply. Table 14-1 shows approximate costs for some types of brick walks.

DRIVES AND PARKING LOTS

Drives and parking lots are usually done by a subcontractor skilled in his particular line of work. The prices given in Table 14-2 will contain the overhead and profit of the subcontractor and thus cover payroll taxes and insurance. To this price the general contractor need only add a small handling charge and profit.

TABLE 14-2
Approximate Cost of Drives and Parking Lots
per Sq Yd

Type of pavement	Cost per sq yd
Penetration macadam — 4″ thick	$1.75
Penetration macadam — 6″ thick	1.95
Bituminous wearing course — 1 1/2″ thick	1.60
Bituminous wearing course — 3″ thick	2.10
Concrete pavement 4″ thick	3.75
Concrete pavement 5″ thick	4.25
Concrete pavement 6″ thick	5.00

As stated for walks it will be necessary to allow for rough and fine grading and curing in addition to the approximate costs shown in Table 14-2.

Bituminous paving for drives and parking lots is widely used perhaps because it is cheaper than concrete and lasts well. When a base course is required, there will be an addition to the cost shown in Table 14-2. For a 3″-base course of gravel add $0.90 per sq yd and for a 6″ course add $1.25 per sq yd.

When a base course is required for bituminous paving, then penetration type macadam is more economical. Where traffic is light this type of pavement is very satisfactory.

Table 14-2 shows approximate costs for some types of pavements.

CURBS

Asphalt curbs are sometimes used in parking lots because the asphalt is much cheaper than concrete. However, concrete is more widely used because of its resistance to the bumping from traffic. Granite is sometimes employed but the cost is much too high for common use.

Table 14-3 gives approximate costs for the several kinds of curbs.

TABLE 14-3
Approximate Cost of Curbs per Lin Ft

Type of curb	Cost per lin ft
Asphalt	$1.50
Concrete 6″ × 18′	4.50
Granite	8.00

CHAPTER FIFTEEN

Mechanical

HEATING

Duct Furnaces

Table 15-1 shows the approximate costs of gas-fired duct furnaces with controls.

The duct work for the furnaces in Table 15-1 priced per lb installed will be:

For galv. steel	—	$1.50 (Assume 1/3 labor)
For aluminum	—	$2.50 (Assume 1/3 labor)

TABLE 15-1
Approximate Cost of Some Gas-Fired Duct Furnaces
Installed

MBH input	Material	Labor Hr
50	$200	5
100	240	13
200	370	21
400	575	30

Note: Labor hours are for skilled labor.

Unit Heaters

Table 15-2 shows the approximate costs of gas-fired unit heaters.

TABLE 15-2
Approximate Cost of Gas-Fired Unit Heaters
Installed (Without Ducts)

Btu output per hour	Material	Labor hr
50,000	$210	9
100,000	310	14
150,000	450	23
200,000	575	29

Note: The above hours are for skilled labor.

Space Heaters

Table 15-3 shows some costs of floor-mounted space heaters complete with fan, controls, and thermostat.

TABLE 15-3
Approximate Cost of Space Heaters Installed

Type	Output	Material	Labor hr
Gas-fired	60 MBH	$320	8
Gas-fired	100 MBH	390	14
Oil-fired	100 MBH	500	50
Oil-fired	150 MBH	700	55

Note: The above hours are for skilled labor.

Steam Heating

The cost of steam heating will vary due to type of fuel used, usually gas or oil, and due to the distribution system and type of radiators used. The complete cost per sq ft of radiation for a one-pipe convector system will probably be in the range of $7.50 to $8.50.

Radiant Heating

Under-the-floor radiant heating with hot water or forced air can currently be installed for $2.00 to $2.50 per sq ft. Radiant electrical heating installations will cost in the neighborhood of $20 per linear foot of baseboard.

COOLING

Central Cooling

For central air conditioning with evaporating condensers, or cooling towers, in sizes above 25 tons, the cost per ton will vary from $1,000 to $1,500. This item is usually a subcontract item.

Unit Air Conditioners

The prices shown in Table 15-4 are for unit air conditioners installed on a subcontract basis.

TABLE 15-4
Approximate Cost of Unit Air Conditioners Installed

Type	Size	Cost
Window units	1/2 hp	$350
Window units	3/4 hp	400
Window units	1 hp	450
Window units	1 1/2 hp	550
Floor type	to 10 tons	400
Water cooled	above 10 to 30 tons	300

PLUMBING

Table 15-5 will serve to give some idea of the cost of plumbing fixtures installed. These costs will vary widely due to quality of material used and differing labor scales.

Table 15-6 gives some prices for copper piping used to bring in the water service from the street main and the C. I. piping needed to connect to the city sewer main.

TABLE 15-5
Approximate Material Cost and Installation Time
for Plumbing Fixtures

Type of fixture	Fixture cost	Labor hr
Tub, 5' P.E. with shower & curtain	$300	7.5
Shower cabinet, 36" wide	200	9.5
Wall hung lav. 19" X 17" P.E	160	6.0
China lav., 22" X 18", chrome legs	175	6.0
Water closet, vitreous china, stand. type	225	8.5
Medicine cabinets	25 to 100	1.0
Urinals, stall type, vitreous china	240	9.0
Urinals, wall hung	180	8.0
Water heater, 80 gal elec. glass lined	185	5.0
Kitchen sink, cabinet type, two drain boards	185	7.5
Kitchen sink, single bowl P.E. — one drain board	160	7.5

Note: The above times are for plumber with helper and the
prices are for white fixtures — for colored add 15%. These
prices for fixtures include an allowance for roughing-in,
risers, vents, and hot and cold service using copper piping.

    Plumber — $ 9.15
    Helper —   6.30
            $15.45 per labor hour

TABLE 15-6
Approximate Material Cost and Installation Time
for Piping

Type of pipe	Dia. of pipe in in.	Cost per lin ft	Labor hr per lin ft
Copper	3/4"	$1.20	0.10
Copper	1	1.75	0.10
Copper	1 1/4	1.75	0.13
Copper	1 1/2	2.90	0.15
Copper	2	4.60	0.16
Cast iron (caulked joints)	4	2.40	0.30
Cast iron (caulked joints)	6	4.30	0.41
Cast iron (caulked joints)	8	7.50	0.61

Note: The above times are for plumber with helper.

ELECTRICAL

The electrical work on a building is almost always performed by a subcontractor experienced in this kind of work. The subcontractors submit bids to the general contractor, who must be well enough informed to check the bids and to determine whether or not the bids are reasonable in price.

A commonly used approximate method on small- and medium-sized jobs is to base the estimate on an average cost per outlet. There are many variables to consider when estimating electrical wiring and outlets: whether or not conduit or cable is used; the particular code that governs the work; labor rates; union regulations; type and quality of available labor; and the particular job requirements.

To determine the number of outlets count all floor and wall plugs, all switches, and all wall ceiling fixtures.

To the cost of the outlets must be added the lead-in wire, the main switches, main fuse boxes, or circuit breaker boxes.

The electrical work for houses and small- to medium-sized buildings using 110- and 220-volt services would probably show unit prices in the following range:

$6 to $10 per outlet for open work
$10 to $18 per outlet for conduit work
Main fuse box with main switch and circuit breakers—$150 to $350
Lead-in wire 200-amp service—$150 to $200

The final estimate should always be made from the completed plans by someone familiar with electrical work.

ELEVATORS

Elevators are definitely a subcontract item. The opening, which is required by the supplier, is built into the elevator shaft with the specified clearances; then the mechanics will come in, line up and plumb the elevator supports and guides, and install the elevator.

There are many types of lifting devices used for raising both personnel and freight. The installed cost of every type of lifting device varies due to the design requirements. However, some general prices will be shown in Table 15-7 to give an idea of the cost range of this type of equipment.

TABLE 15-7
Approximate Cost of Lifting Devices Installed

Type	Capacity	Cost
Residential	Cab type	$ 4,000—one story
"	"	$ 5,000 — two story
"	Chair lift	$ 1,400 — per story
Personnel elevators for buildings	2,000 lb 100 f p m	$20,000 — first 2 floors plus $1,500 ae. additional
"	4,000 lb 100 f p m	$25,000 — first 2 floors plus $1,500 ea. additional
Freight	4,000 lb	$15,000 to $20,000 — first 2 floors plus $2,500 ea. additional
Freight	10,000 lb	$22,000 to $32,000 — first 2 floors plus $2,500 ea. additional
Escalators	—	$35,000 per unit per story
4' moving ramps	—	$ 600 per lin ft
Dumbwaiters (hand operated)	—	$ 2,200 plus $300 per stop for all over 2 stops
Dumbwaiters (elec. operated)	—	$ 3,300 plus $800 per stop for all over 2 stops

SPRINKLER SYSTEMS

There are several types of sprinkler systems for which the prices will vary from $40 to $60 per sprinkler head when as many as 200 heads are used in the system. When 100 to 200 heads are used, $5.00 to $10.00 will have to be added to the cost of each head.

The cost of valves and operating devices will have to be added to the costs. This is a subcontract item and the general contractor's handling cost plus a profit should be added to the costs obtained by this method.

DISTRIBUTION OF BUILDING COSTS

Tables 15-8 and 15-9 show some approximate distribution of costs in various types of buildings. Table 15-8 shows the distribution of the cost between material, labor, equipment, overhead, and profit. Table 15-9 shows the distribution of the cost between the several mechanical items.

TABLE 15-8
Average Distribution of Building Costs as
Percentages of Total Cost

Type of construction	Percentage of total costs				
	Material	Labor	Equipment	Overhead	Profit
All types	42%	32%	8%	10%	8%
All buildings	42	33	6	9	10
Commercial	39	34	7	10	10
Manufacturing	50	25	5	9	11
Residential	43	32	4	9	12

TABLE 15-9
Average Mechanical Cost as Percentages
of Total Cost

Type of Building	Plumbing	Heating and cooling	Electrical
Residential housing	10.0%	12.0%	7.5%
Housing projects—low rent	10.0	5.5	6.0
Apartments	9.5	8.5	7.0
Dormitories	8.0	12.5	8.0
Motels	9.0	12.0	9.0
College classrooms	6.0	17.0	10.0
College science labs	8.0	17.5	10.5
Hospitals	9.0	16.0	11.5
Schools — secondary	7.0	12.0	10.0
Office buildings	5.5	13.5	10.5
Warehouses	5.0	6.0	7.5
Factories	6.0	14.0	11.0

Every architect and engineer who prepares many bids will have in his files the square foot or cubic foot costs of all of the buildings that he has let to contract. These prices are of real value in estimating similar work. From such information the architect or engineer can quickly determine if the client is expecting a larger building or a more complicated structure than his building fund will permit.

Such figures are also of help to the contractor. There are several ways that he can check his detailed estimate, such as by comparing it to a like structure in total cost or a like structure in cost per square or cubic foot.

Table 15-10 will show some average sq ft costs. It should be remembered that such costs are only a guide because so many considerations can change these costs considerably.

TABLE 15-10
Average Square Foot Cost in Dollars

Type of building	Average low	Average high
Residential housing	16	24
Housing projects — low rent	14	22
Apartments	13	20
Dormitories	21	28
Motels	18	24
College classrooms	24	37
College science labs	26	40
Hospitals	33	45
Schools — secondary	18	28
Office buildings	21	33
Warehouses	8	14
Factories	10	17

Note: In assembling the above sq ft costs, very cheap and very expensive construction were omitted and only the average low and high figures were used.

AVERAGE RENTAL RATES FOR CONSTRUCTION EQUIPMENT

The following tables are excerpts from the "22nd Edition (1971) Compilation of Nationally Averaged Rental Rates" and are here reproduced with the permission of its author, the Associated Equipment Distributors of Chicago, Ill. This material was prepared in the latter part of 1970.

A few statements from the 22nd Edition are quoted below to explain the purpose that the average rental rates are supposed to serve.

"What it is: This is a publication of nationally averaged rental rates compiled from thousands of rental charges by A. E. D. member equipment distributors throughout the United States. The figures shown are purely statistical averages—no more—no less.

"The rental figures in this compilation in addition to being National Averages, reflect an averaging of age, condition and operating efficiency of the equipment. New and practically new equipment rents for substantially more than the average reflected in this publication.

"What it is not: It is not a suggested rental rates book. It does not establish maximum or minimum levels for rental rates and the rates are not typical for any one section of the country.

"General industry practices:

"It is the general practice in the industry to base rates upon one shift of 8 hours per day, 40 hours per week, or 176 hours per month."

The 22nd Edition goes on to state that "the lessor usually bears the cost of repairs due to normal wear and tear on non-tractor equipment, and the lessee bears all other costs."

In the case of tractor equipment or rubber-tired hauling equipment the lessee is usually required to pay for the tire wear at least and sometimes for all costs of repair.

In the case of cranes or shovels the lessor usually bears the cost of repairs due to normal wear and tear.

In no case does the average rental rate in this publication include the cost of operator or fuel and lubricants.

The lessee is required to return the rented equipment in as good a condition of repair as he received it.

All rental charges are f. o. b. the lessor's warehouse or shipping point.

No insurance, license, sale or use taxes are included in the rates.

Air compressors—portable
One, Two or Multiple Stage—2 or 4 Wheel Mounting

	free air delivered at 100 lbs.—cfm	per month	per week	per day
Diesel Engine Powered	74-78	$ 183.00	$ 62.25	$ 19.25
	115	*	*	*
	125	241.00	82.00	23.75
	150	256.00	89.50	30.50
	160	289.00	98.00	31.75
	175	294.00	102.00	32.50
	185-190	333.00	113.00	40.00
	210	411.00	141.00	41.75
	250	459.00	157.00	48.75
	315	537.00	185.00	60.25
	365	613.00	208.00	68.00
	450	760.00	253.00	82.50
	600	886.00	305.00	100.00
	750	972.00	327.00	113.00
	900	1300.00	444.00	147.00
	1200	1777.00	601.00	198.00
	1500	2062.00	725.00	236.00

Portable Compressors are ordinarily those consisting of compressors, power unit, and receiver with or without running gear but so mounted that they may be moved as a unit.

Wrenches, impact

	per month	per week	per day

Standard type (without socket)

Capacity—bolt size, inches

	per month	per week	per day
¼	$ 61.50	$21.50	$ 6.75
⅜	*	*	*
⅝	65.25	22.25	7.50
¾ Standard duty	84.75	30.00	9.65
¾ Heavy duty	93.00	32.75	11.25
1-1¼	112.00	37.50	12.50
1½	143.00	49.25	16.50
1¾	*	*	*
2	162.00	55.25	18.00
4	*	*	*

Controlled torque type (without socket)

capacity high strength bolt	torque range	per month	per week	per day
⅜	20-90 ft. lbs.	$112.00	$39.75	$13.25
⅞	300-550 ft. lbs.	171.00	58.25	19.25

* insufficient information received

248   *Appendix A*

Air tool accessories

		per month	per week	per day

Hose, air

diameter in.	length ft.	per month	per week	per day
1/2	50	$ 8.55	$ 3.45	$ 1.50
5/8	50	10.50	4.30	1.80
3/4	50	12.00	4.90	2.00
1	25	15.00	5.80	2.35
1	50	15.50	5.85	2.35
1 1/4	25	28.50	9.75	3.40
1 1/4	50	35.25	13.00	4.70
1 1/2	25	28.25	10.75	4.05
1 1/2	50	36.75	13.50	4.90
2	25	43.25	15.75	5.75
2	50	56.75	20.50	7.55

Whip hose—

	per month	per week	per day
1/2" x 10'	$ 5.30	$ 2.15	$ 1.00
3/4" x 10'	*	*	*
1" x 10'	*	*	*

Compactors—vibratory plate type
Manually guided

weight (lbs.) from	to	plate width (in.) from	to	per month	per week	per day

Gasoline powered—single unit

from	to	from	to	per month	per week	per day
—	150	9	19	$163.00	$ 54.25	$18.00
151	280	12	24	169.00	62.00	19.25
281	360	18	30	200.00	68.00	22.00
361	500	20	36	246.00	81.50	22.50
521	800	24	32	317.00	111.00	35.25
801	1000	20	36	371.00	127.00	42.25

Buckets—concrete

capacity—cu. yds.	per month	per week	per day

Bottom dump—manually operated

capacity—cu. yds.	per month	per week	per day
Under 1/2	$ 54.50	$ 19.75	$ 8.15
1/2	64.25	23.50	8.60
3/4	81.25	30.25	10.75
1	94.50	33.75	12.00
1 1/2	122.00	44.50	15.50

* insufficient information received

Buckets—concrete

capacity—cu. yds.	per month	per week	per day
2	141.00	52.75	17.00
3	204.00	*	*
4	*	*	*

Laydown

	per month	per week	per day
Under ½	$ *	$ *	$ *
½	77.00	27.50	10.75
¾	87.00	31.75	11.25
1	97.50	36.75	13.75
1½	120.00	43.25	15.50
2	152.00	52.25	18.50
3	*	*	*
4	*	*	*

Buggies—concrete, manually operated
Pneumatic tired

	per month	per week	per day
All sizes	$ 36.50	$ 13.75	$ 5.25

Buggies—concrete, power operated
Pneumatic tired

concrete capacity—cu. ft.

from & not including	to & including	per month	per week	per day
Operator walking type				
—	10	$178.00	$ 61.50	$20.75
10	12	196.00	66.75	22.00
Operator riding type				
—	10	$191.00	$ 63.00	$21.00
10	14	204.00	70.50	24.00
14	16	210.00	76.50	24.25
16	18	243.00	82.00	26.75
22	28	303.00	103.00	32.50

Buggies—mortar, manually operated
Cushion tired

	per month	per week	per day
6 cu. ft.	$ 37.50	$ 13.00	$ 5.30

* insufficient information received

Cranes—lifting
Cable controlled

capacity—tons	per month	per week	per day

All rates for cranes and draglines are with factory standard length booms and do not include extensions, boom jibs, hook blocks or other special accessories. When longer booms are furnished, extra charges are made for lengths over standard. Crawler crane rates are based on cranes with standard crawlers. Crawler crane ratings are based on capacities which do not exceed 75 per cent of tipping load without outriggers. Truck mounted and wheel mounted ratings are based on capacities which do not exceed 85 per cent of tipping load with or without outriggers.

Crawler mounted, gasoline engine powered (no bucket)

capacity—tons	per month	per week	per day
Thru 12½	$ 917.00	$ 333.00	$117.00
15	1475.00	463.00	133.00
20	*	*	*
25	*	*	*
30	2225.00	800.00	225.00
35	*	*	*
45	2600.00†	*	*

Crawler mounted, diesel engine powered (no bucket)

capacity—tons	per month	per week	per day
Thru 12½	$1338.00	$ 389.00	$125.00
15	1559.00	511.00	*
20	1803.00	546.00	189.00
25	1940.00	689.00	250.00
30	2206.00	761.00	256.00
35	2569.00	809.00	*
45	2808.00	905.00	300.00
60	3307.00	1164.00	346.00
90	3914.00	*	*
120	4413.00	1310.00	*
Over 120	6040.00	1967.00	*

Truck mounted, gasoline engine powered (no bucket)

capacity—tons	per month	per week	per day
Thru 12½	$ *	$ *	$ *
15	1539.00	482.00	130.00
20	1683.00	*	*
25	1960.00	625.00	213.00
30	2271.00	850.00	236.00
35	2664.00	890.00	263.00
45	2934.00	1025.00	362.00
60	3358.00	1125.00	462.00

Truck mounted, diesel engine powered (no bucket)

capacity—tons	per month	per week	per day
Thru 12½	$ *	$ *	$ *
15	*	*	*
20	*	*	*
25	2286.00	763.00	235.00
30	2575.00	800.00	268.00

* insufficient information received   † This average based on limited returns.

Cranes—lifting
Cable controlled

capacity—tons	per month	per week	per day
35	2727.00	921.00	294.00
45	3148.00	936.00	337.00
60	3645.00	1210.00	393.00
90	4259.00	1445.00	437.00
120	5550.00	*	*

Cranes—lifting
Hydraulic Telescoping—Swinging Boom
Truck or carrier mounted (integral)

capacity—tons	per month	per week	per day
Gasoline engine powered			
5	$1013.00	$313.00	$ 90.00
6	1020.00	435.00	114.00
10	1368.00	470.00	118.00
15	1745.00	597.00	186.00
20	2125.00	687.00	221.00
25	2538.00	824.00	275.00
35-40	2933.00	*	*
Diesel engine powered			
12-12½	$1314.00	$438.00	$153.00
15	1583.00	531.00	183.00
17½	1767.00	592.00	190.00
18	1780.00	600.00	*
25	2665.00	856.00	261.00
30	2763.00	924.00	317.00
45	3450.00	*	*
55	3738.00	*	*
65	4060.00	*	*

Shovels—power
Hydraulic

bucket capacity (cu. yds.)	per month	per wk	per day
Crawler mounted, diesel engine powered			
⅞	$1867.00	$ *	$ *
1	*	*	*
1¼	2167.00	*	*
2¼	*	*	*
Self-propelled, gasoline engine powered			
1¼	$1960.00	$ 628.00	$206.00
1¼	$2075.00	$ 633.00	$211.00

* insufficient information received

252   *Appendix A*

Shovels—pull (Backhoes)
Mechanical

bucket capacity (cu. yds.)	per month	per week	per day
Crawler mounted, diesel engine powered			
3/8	$ *	$ *	$ *
1/2-5/8	1117.00	*	*
3/4	2091.00	700.00	233.00
1	2400.00	*	*
1 1/4	2967.00	*	*
1 1/2	3350.00	*	*
1 3/4	*	*	*
2	*	*	*

Shovels—pull (Backhoes)
Hydraulic

bucket capacity (cu. yds.)	per month	per week	per day
Crawler mounted, gasoline engine powered			
5/8	$1188.00†	$ *	$ *
3/4	1404.00	488.00	150.00
7/8	1858.00†	*	*
1	1877.00†	*	*
Crawler mounted, diesel engine powered			
1/2	$1530.00	$ 454.00	$161.00
5/8	1542.00	514.00	174.00
3/4	1802.00	589.00	190.00
7/8	1869.00	620.00	208.00
1	2350.00	802.00	283.00
1 1/4	3088.00	973.00	333.00
1 1/2	3268.00	1098.00	371.00
2	3861.00	1397.00	495.00
2 1/2	5565.00	1800.00	623.00

Hoists—electric powered
Including suitable starting equipment

horsepower from & not including	to & including		per month
Single drum			
—	12		$ *
12	17		*

---

*   insufficient information received   † This average based on limited returns.

Hoists—electric powered
Including suitable starting equipment

horsepower from & not including	to & including	per month
17	27	151.00
27	37	*
37	47	*
47	57	232.00
57	67	*
67	92	*
92	112	349.00
112	137	430.00

Double drum

12	17	$207.00
17	27	*
27	37	*
37	47	317.00
47	57	*
57	67	353.00
67	92	450.00
92	112	593.00
112	137	*
137	177	*

Three drum

67	82	$525.00
82	112	673.00

Hoists—gasoline powered

horsepower from & not including	to & including	per month	per week	per day
Single drum				
—	8	$ 73.25	$ 26.25	$ 9.00
8	14	75.50	30.75	10.25
14	22	99.00	33.25	12.00
22	29	131.00	*	*
29	42	167.00	57.25	21.25
42	55	217.00	*	*
55	65	238.00	*	*
65	78	*	*	*
78	92	*	*	*
92	112	*	*	*

* insufficient information received

Hoists—gasoline powered

from & not including	horsepower to & including	per month	per week	per day

Double drum

from & not including	to & including	per month	per week	per day
—	22	$ *	$ *	$ *
22	28	*	*	*
28	45	205.00	72.00	26.50
45	55	228.00	89.50	31.50
55	65	287.00	*	*
65	75	*	*	*
75	90	373.00	*	*
90	110	407.00	*	*

Three drum

from & not including	to & including	per month	per week	per day
—	42	$ *	$ *	$ *
42	57	*	*	*
57	72	*	*	*
72	87	442.00	*	*
87	112	575.00	*	*

Towers—steel, self-erecting, portable
Gasoline engine powered—with hoist

	per month	per week	per day
Extra heavy duty—3000 lbs. to 5000 lbs. capacity—single platform			
Up to & including 25 ft. (manual)	$502.00	$178.00	$48.25
Up to & including 25 ft. (automatic)	479.00	172.00	*
Extra 10 ft. section	26.00	11.25	4.85
Extra 6½ ft. section	*	*	*
¾ yd. concrete bucket	85.25	30.75	13.50
Heavy duty—2000 lbs. to 3000 lbs. capacity—single platform			
Up to & including 25 ft. (manual)	$388.00	$135.00	$40.75
Up to & including 25 ft. (automatic)	393.00	139.00	51.75
Extra 10 ft. section	21.75	8.95	3.40
Extra 5 ft. section	*	*	*
17 cu. ft. concrete bucket	71.25	25.00	*
14 cu. ft. concrete bucket	49.00	19.75	*

Hammers—pile, diesel

manufacturer	model	per month	per week	per day
Delmag	D5	$ *	$ *	$ *
	D12	1586.00	*	*

* insufficient information received

Hammers—pile, diesel

manufacturer	model	per month	per week	per day
Link Belt	D22	*	*	*
	105	*	*	*
	180	*	*	*
	312	*	*	*
	440	1828.00†	*	*
	520	*	*	*
MKT—McKiernan Terry	DE20	1357.00†	557.00	*
	DE30	1560.00†	640.00	*
	DA35	1773.00†	665.00	*
	DE40	2048.00†	823.00	*

Trade names are mentioned to indicate size. Hammers of manufacturers not mentioned are subject to the same rates as similar sizes and types of names mentioned.

Hammers—pile, drop

pounds from & not including	pounds to & including	per month	per week	per day
—	1,500	$ *	$ *	$ *
1,500	2,250	*	*	*
2,250	3,000	*	*	*
3,000	4,000	*	*	*

Hammers—pile, steam

manufacturer	model	per month	per week
MKT—McKiernan Terry Single Acting	S-3	$ *	$ *
	S-5	*	*
	S-8	*	*
	S-10	*	*
	S-14	*	*
Double Acting	0	*	*
	1	143.00	64.00
	2	188.00	86.00
	3	258.00	104.00
	5	323.00	127.00
	6	415.00	173.00
	6.5	*	*
	7	478.00	192.00

* insufficient information received   † This average based on limited returns.

Hammers—pile, steam (continued)

manufacturer	model	per month	per week
	9B3	745.00	288.00
	10B3	904.00	384.00
	11B3	1203.00	490.00
	C3	*	*
	C5	*	*
	C8	*	*
Vulcan Iron Works, Inc.			
Single acting	3	$ 375.00	$ 163.00
	2	504.00	193.00
	1	620.00	235.00
	0	1025.00	428.00
	OR	1193.00†	530.00
Differential acting	DGH-100	$ 265.00	$ 98.00
	DGH-900	438.00	187.00
	18C	*	*
	30C	631.00	263.00
	50C	813.00	327.00
	80C	1291.00†	568.00
	140C	2100.00†	875.00
	200C	2833.00†	1163.00
	400C	*	*
MKT-McKiernan Terry *(Vibratory)*	V-10	6800.00†	2650.00

Trade names are mentioned to indicate size. Hammers of manufacturers not mentioned are generally rented at the same rates as similar sizes and types of names mentioned.

Hammer leads and attachments

	per month	per week
40 ft.—48 ft. pile driver leads	$ 235.00	$ *
10 ft. Center Section	55.00	*
20 ft. Center Section	100.00	*
Pile driver pants	*	*
H-beam helmets	107.00	31.75
Sheet piling helmets	119.00	37.50
Flat or dished anvils or caps	*	*
Pile shackles	12.50	*
Lubricators with quick acting valves	29.00	*
400 to 750 gallon air receivers	*	*

* insufficient information received † This average based on limited returns.

Crawler tractors—without attachments

from & including	horsepower to & including	per month	per week

Gear drive—diesel engine
Drawbar horsepower

from & including	to & including	per month	per week
20	25	$ *	$ *
26	35	709.00	*
36	44	900.00	300.00
45	59	998.00	*
60	71	*	*
72	79	1252.00	*
80	91	*	*
92	115	1605.00	*
116	—	*	*

Torque converter drive—diesel engine
Net engine horsepower

from & including	to & including	per month	per week
115	144	$1650.00	$ *
320	360	3567.00†	*

Crawler tractors—with bulldozer

from & including	horsepower to & including	per month	per week

Gear drive—diesel engine
Drawbar horsepower

from & including	to & including	per month	per week
20	25	$ *	$ *
26	35	663.00	233.00
36	44	874.00	287.00
45	59	999.00	348.00
60	71	*	*
72	79	1386.00	481.00
80	91	$ *	$ *
92	115	1637.00	551.00
116	131	1998.00	678.00
132	150	2233.00†	*
151	170	2447.00	818.00
171	200	*	*
201	256	3133.00	996.00
257	—	3625.00	*

---

* insufficient information received   † This average based on limited returns.

Crawler tractors—with loader (shovel) front

Front end load and dump—complete with manufacturer's standard general purpose bucket, S.A.E. rated capacity based on material weighing 3,000 lbs. per cubic yard.

S.A.E. rated capacity, cu. yds. to & including	per month	per week	per day
Diesel Engine—Torque Converter, Power Shift			
1	$1019.00	$344.00	$109.00
1¼	1091.00†	380.00	120.00
1½	1316.00	423.00	126.00
1¾	1604.00	561.00	169.00
2	1811.00	593.00	*
2¼	1885.00	680.00	*
2½	2362.00	795.00	189.00
2¾	2540.00	881.00	282.00
5	4358.00	*	*

Motor scrapers and hauling units
Diesel engine powered

rated engine horsepower		cu. yd. capacity without sideboards						
		struck		heaped				
from & inc.	to & inc.	from & inc.	to & inc.	from & inc.	to & inc.	per month	per week	per day
2 wheel tractor with 2 wheel scraper (single engine drive)								
90 - 161	5.5 - 9		7	- 11		$ *	$ *	$ *
162 - 200	7.5 - 12.5		10.5 - 18			2017.00	650.00	*
200 - 230	12 - 13		16 - 18			*	*	*
200 - 230	14 - 18		18 - 25			2567.00	875.00	*
231 - 300	12 - 18		18 - 25			2689.00	954.00	321.00
300 - 340	19 - 24		27 - 33			3970.00	1313.00	345.00

Note: rates do not include additional charges for tire wear.

2 wheel tractor with 2 wheel scraper (single engine drive)								
341 - 430	20 - 24		27 - 33			$4398.00	$1390.00	$407.00
431 - 475	24 - 25		32 - 34			4398.00	1390.00	407.00
475 - 520	28 - 32		38 - 44			5796.00	1889.00	*

4 wheel tractor with 2 wheel scraper (single engine drive)								
141 - 190	10 - 14		12 - 18			$1450.00	$ *	$ *
191 - 220	12 - 14		16 - 22			*	*	*
221 - 300	14 - 18		20 - 25			3313.00	*	*
300 - 350	18 - 27		25 - 35			*	*	*

* insufficient information received   † This average based on limited returns.

Motor scrapers and hauling units (continued)
Diesel engine powered

rated engine horsepower		cu. yd. capacity without sideboards						
		struck		heaped				
from & inc.	to & inc.	from & inc.	to & inc.	from & inc.	to & inc.	per month	per week	per day

2 wheel tractor with 2 wheel scraper—two engine drive, 1 driving tractor wheels, 1 driving scraper wheels

from & inc.	to & inc.	from & inc.	to & inc.	from & inc.	to & inc.	per month	per week	per day
251 - 320	— 14			— 20		$3376.00	$1191.00	$425.00
500 - 600	16 - 18			21 - 25		*	*	*
561 - 660	19 - 24			26 - 32		5250.00	1707.00	558.00
661 - 900	32 - 33			40 - 44		*	*	*

Note: rates do not include additional charges for tire wear.

Motor scrapers and hauling units—self loading
Diesel engine powered

rated engine horsepower					
from & inc.	to & inc.	rated capacity (cubic yards)	per month	per week	per day

2 wheel tractor with 2 wheel scraper

from & inc.	to & inc.	rated capacity (cubic yards)	per month	per week	per day
121 - 144		8 - 9	$1898.00	$ 685.00	$217.00
145 - 190		10 - 12	2176.00	752.00	233.00
191 - 288		13 - 19	2760.00	945.00	307.00
250 - 350		20 - 27	3698.00	1254.00	401.00
400 - 500		28 - 32	5110.00	1608.00	528.00

4 wheel tractor with 2 wheel scraper

from & inc.	to & inc.	rated capacity (cubic yards)	per month	per week	per day
121 - 144		8 - 9	$1728.00	$ 600.00	$175.00
145 - 190		9.5- 12	1944.00	612.00	189.00

Note: rates do not include additional charges for tire wear.

---

* insufficient information received

Trucks—rear dump, off-the-highway
Single drive axle

from & not including	to & including	per month
body capacity—tons		
7.5	10	$ *
10	15	1467.00
15	20	*
20	23	2645.00
23	25	*
25	30	*
30	35	3500.00
35	40	3658.00
40	50	5545.00

Note: rates do not include additional charges for tire wear.

Conveyors—belt, portable
Gasoline or electric powered

width of belt from & not including (in.)	to & including (in.)	conveyor length C. to C. from & not including (ft.)	to & including (ft.)	per month	per week	per day
—	12	—	20	$ *	$ *	$ *
		20	26	*	*	*
		26	30	*	*	*
12	18	—	20	$ *	$ *	$ *
		20	26	257.00	90.50	31.50
		26	30	*	*	*
		30	36	*	*	*
		36	46	448.00	166.00	61.50
		46	56	542.00	179.00	67.00
18	24	26	30	$ *	$ *	$ *
		30	36	*	*	*
		36	46	*	*	*
		46	56	*	*	*

* insufficient information received

The following table is a reprint from "Building Construction Cost Data-1971" by Robert Snow Means Company, Inc., who are engineers and estimators located in Duxbury, Mass.

"Labor rates used in this edition are as listed below for 1971. They are averages of the 30 largest cities in the U.S. as reported by the U.S. Department of Labor and are substantially the same as listed by *Engineering News-Record*. The rates have been rounded out to the nearest 5¢ and include fringe benefits but do not include insurance or taxes."

Trade	1971	1970	1969	1968	1967
Common Building Labor	$6.05	$5.00	$4.55	$4.10	$3.85
Skilled Average	7.95	6.85	6.05	5.50	5.15
Helpers Average	6.15	5.15	4.65	4.20	4.00
Foremen (usually 35¢ over trade)	8.30	7.20	6.40	5.85	5.50
Bricklayers	8.40	7.15	6.40	5.85	5.55
Bricklayers Helpers	6.20	5.20	4.70	4.30	4.05
Carpenters	7.85	6.95	6.15	5.40	5.10
Cement Finishers	7.70	6.75	5.90	5.30	5.05
Electricians	8.75	7.50	6.45	5.95	5.60
Glaziers	7.35	6.25	5.50	5.10	4.75
Hoist Engineers	7.95	7.05	5.90	5.40	5.10
Lathers	7.80	6.60	5.95	5.45	5.20
Marble & Terrazzo Workers	7.70	6.45	5.60	5.25	5.05
Painters, Ordinary	7.25	6.20	5.45	5.05	4.75
Painters, Structural Steel	7.45	6.50	5.80	5.30	4.95
Paperhangers	7.35	6.30	5.60	5.15	4.75
Plasterers	7.75	6.60	5.95	5.50	5.15
Plasterers Helpers	6.50	5.30	4.85	4.45	4.15
Plumbers	9.15	7.75	6.90	6.15	5.75
Power Shovel or Crane Operator	8.20	7.20	6.20	5.65	5.35
Rodmen (Reinforcing)	8.35	7.30	6.35	5.80	5.45
Roofers, Composition	7.45	6.30	5.55	5.05	4.75
Roofers, Tile & Slate	7.50	6.35	5.60	5.10	4.85
Roofers Helpers (Composition)	5.75	4.75	4.45	4.00	3.75
Steamfitters	9.15	7.70	6.90	6.10	5.70
Sprinkler Installers	9.05	7.70	6.90	6.10	5.70
Structural Steel Workers	8.55	7.45	6.45	5.90	5.55
Tile Layers (Floor)	7.55	6.50	5.60	5.20	4.90
Tile Layers Helpers	6.30	5.25	4.80	4.35	4.15
Truck Drivers	6.20	5.15	4.60	4.30	3.95
Welders, Structural Steel	8.35	7.15	6.35	5.80	5.45

Appendix C

WORKMEN'S COMPENSATION INSURANCE–BASE RATES–JULY 1, 1969

The base rates on Workmen's Compensation for construction given in the following table were reprinted by permission of the editors of *Engineering News Record* (September 18, 1969, p. 133, copyright, McGraw-Hill, Inc. All rights reserved).

The rates given in the table are the basic manual rates per $100 of payroll applicable to each classification.

For each employer whose premium volume is large enough for an experience rating (generally $750 annual premium) the premiums calculated at these manual rates must be adjusted by an experience modification based on the risk's own past experience.

Base Rates on Compensation Insurance

Rate/$100 payroll. Compiled by Herbert L. Jamison & Co., Insurance Advisers and Auditors, New York, N. Y.

Classification of Work	Ala.	Alaska	Ariz.	Ark.	Cal.	Colo.	Conn.	Del.	D.C.	Fla.	Ga.	Hawaii	Ida.
Carpentry—1, 2 Family Residence	$2.08	$3.03	$5.23	$4.45	$4.88	$3.44	$4.60	$2.20	$3.26	$4.99	$3.15	$4.77	$3.38
Carpentry—3 Stories or less	2.24	3.03	5.23	5.62	4.88	4.55	4.60	2.20	3.48	4.99	3.15	5.22	3.38
Carpentry—Interior Cab WK	1.01	1.87	5.23	2.82	4.88	2.07	2.11	2.20	2.66	3.57	1.51	3.47	2.00
Carpentry—General	2.54	3.60	5.23	5.86	8.66	4.63	4.60	2.20	5.28	7.03	4.09	12.07	3.38
Chimney—Construction Brick Con	7.15	4.10	7.63	22.16	19.13	11.71	14.90	2.00	18.22	22.45	5.08	18.60	14.00
Concrete Work—Bridges Culverts—C	3.92	4.48	10.51	10.44	9.98	6.69	10.06	2.00	11.13	7.42	3.90	7.51	3.19
Concrete Work Dwellings 1-2 Fam	1.34	5.03	10.51	2.98	2.79	2.99	2.06	2.00	4.28	4.91	1.06	4.30	1.78
Concrete Work—N.O.C.	1.86	4.10	7.22	3.96	7.36	4.96	3.73	2.00	6.45	8.62	2.31	7.48	2.45
Concrete or Cement Work—Floors, Sidewalks	1.18	2.98	5.36	2.73	2.79	2.93	2.93	0.85	2.93	4.57	1.89	2.72	1.68
Electrical Wiring—Inside	1.23	1.57	3.38	3.34	2.84	1.33	1.35	0.77	1.94	2.82	1.48	3.39	1.90
Excavation Earth N.O.C.	2.00	3.72	3.38	4.80	3.61	3.87	4.96	3.75	4.39	6.40	2.94	13.72	8.26
Excavation—Rock	3.45	3.67	7.24	6.14	8.35	5.21	4.37	3.75	13.57	11.56	5.15	11.55	3.66
Glaziers	2.70	3.04	4.72	5.60	5.31	4.07	4.30	1.65	3.66	5.15	4.46	6.61	2.48
Insulation Work	1.72	4.61	5.23	2.31	5.33	2.90	3.73	2.20	3.20	4.88	1.83	3.36	3.54
Lathing	1.29	2.93	2.75	3.61	4.20	2.07	1.77	1.10	2.56	2.71	1.65	3.64	2.27
Masonry	1.34	4.07	5.32	2.98	5.63	3.26	2.81	1.80	2.87	3.45	1.52	4.30	1.78
Painting and Decorating	2.32	4.40	4.73	2.74	5.31	2.19	3.28	2.25	2.61	5.18	2.56	3.64	2.16
Pile Driving	8.92	9.27	15.58	14.85	14.30	12.41	3.36	6.95	16.55	12.43	5.22	12.02	13.11
Plastering	1.72	2.00	4.67	2.31	7.06	2.90	3.73	1.10	3.20	4.79	1.83	3.36	3.54
Plumbing	1.35	2.02	3.10	3.06	3.08	1.97	1.82	0.81	2.44	3.54	1.83	2.96	1.39
Roofing	3.52	7.58	9.52	12.07	12.61	7.45	9.24	3.00	8.09	10.64	4.81	16.23	7.23
Sheet Metal Work—Erection—Installation and Repair	1.70	1.95	4.10	3.35	3.29	2.83	3.16	3.00	3.22	4.47	2.16	4.78	2.27
Steel Erection—Doors and Sash	2.08	6.05	2.23	4.26	4.35	3.30	3.58	3.30	3.11	4.50	1.86	3.38	3.22
Steel Erection—Interior Ornament	2.08	6.05	2.23	4.26	4.35	3.30	3.58	3.30	3.11	4.50	1.86	3.38	3.22
Steel Erection—Structure	3.85	13.08	29.26	9.73	16.99	14.69	13.53	3.30	27.45	15.78	8.39	24.90	4.34
Steel Erection—Dwelling 2 Stories	3.29	7.91	10.85	7.33	13.28	6.22	7.08	3.30	8.91	9.72	5.43	11.74	4.34
Steel Erection—N.O.C.	3.52	12.90	10.85	8.62	12.34	9.74	11.86	3.30	24.12	10.21	7.28	24.03	4.34
Tile Work—Interior	1.04	0.85	1.83	2.25	2.73	1.91	1.72	0.78	4.55	3.42	1.11	2.68	1.42
Timekeepers and Watchmen	1.89	4.89	2.05	4.20	2.51	3.56	3.99	B	3.88	4.76	2.72	4.16	3.54
Waterproofing Interior (Brush)	2.32	4.40	4.73	2.74	5.31	2.19	3.28	2.25	2.61	5.18	2.56	3.64	2.16
Waterproofing—Trowel Interior	1.72	2.00	4.67	2.31	7.06	2.90	3.73	1.10	3.20	4.79	1.83	3.36	3.54
Waterproofing—Trowel Exterior	1.34	4.07	5.32	2.98	5.63	3.26	2.81	1.80	2.87	3.45	1.52	4.30	1.78
Waterproofing Pressure Gun	1.86	4.10	7.22	3.96	7.36	4.96	3.73	2.00	6.45	8.62	2.31	7.48	2.45
Wrecking	8.52	11.04	22.02	35.85	20.40	24.59	17.73	6.40	17.24	29.56	10.98	35.72	24.90

Classification of Work	Mo.	Mont.	Neb.	(C) Nev.	N.H.	N.J.	N.M.	N.Y.	N.C.	(C) N.D.	(C) Ohio	Okla.	Ore.
Carpentry—1, 2 Family Residence	$3.16	$4.35	$2.32	$5.96	$2.60	$5.60	$3.54	$3.10	$2.44	$3.26	$1.88	$5.35	$5.23
Carpentry—3 Stories or less	3.16	4.35	2.58	5.96	2.60	2.98	3.54	3.10	2.43	3.26	2.43	6.70	5.23
Carpentry—Interior Cab WK	2.04	1.80	1.08	2.47	1.25	3.06	2.16	4.30	2.43	3.26	1.21	3.90	5.23
Carpentry—General	4.50	4.44	4.12	6.08	2.60	6.64	5.55	4.30	3.58	6.10	2.93	6.70	6.66
Chimney—Construction Brick Con	13.04	16.48	8.30	22.58	9.57	14.20	11.42	20.20	10.52	15.83	7.76	22.99	16.55
Concrete Work—Bridges Culverts—C	4.39	7.21	3.55	9.88	4.74	6.91	11.00	6.90	4.47	4.86	3.62	15.80	8.52
Concrete Work Dwellings 1-2 Fam	2.33	3.80	1.44	5.21	1.80	3.67	1.60	4.60	1.34	2.13	1.84	2.28	4.18
Concrete Work—N.O.C.	4.96	4.94	3.19	6.77	3.36	6.91	4.54	4.60	3.03	4.89	2.72	7.78	4.38
Concrete or Cement Work—Floors, Sidewalks	2.44	2.19	1.41	3.00	1.65	3.33	2.27	2.40	1.35	1.79	1.24	3.20	3.40
Electrical Wiring—Inside	1.99	1.22	1.39	1.67	1.21	3.00	2.48	2.20	1.46	1.66	1.28	3.11	3.06
Excavation Earth N.O.C.	4.10	5.70	2.02	7.81	3.72	5.06	2.88	3.90	3.19	3.45	2.48	6.82	7.41
Excavation—Rock	4.93	5.28	6.29	7.23	5.12	8.47	5.35	3.40	5.68	4.29	5.22	6.65	6.10
Glaziers	3.73	5.47	2.93	7.49	3.56	6.33	4.54	4.40	2.93	4.39	2.23	5.75	2.74
Insulation Work	2.90	5.06	1.49	6.93	1.87	3.91	3.24	3.20	3.73	1.44	4.03	6.45	5.24
Lathing	1.67	2.40	1.43	3.29	1.67	3.46	1.56	1.60	1.70	1.68	1.32	2.91	5.24
Masonry	2.86	3.80	1.44	5.21	1.80	5.31	3.01	3.80	1.34	1.79	1.64	4.06	6.63
Painting and Decorating	3.02	5.88	1.67	8.06	3.09	7.37	2.84	4.20	1.72	2.57	2.15	3.93	6.45
Pile Driving	7.25	14.42	6.81	19.76	4.70	9.43	13.31	6.70	5.45	10.11	4.75	15.69	15.92
Plastering	2.47	5.06	1.49	6.93	1.87	3.46	3.24	4.80	1.80	1.79	1.44	3.62	3.20
Plumbing	2.74	2.83	1.30	3.88	1.74	3.46	2.24	2.50	1.46	2.32	1.44	3.56	3.37
Roofing	7.41	10.60	6.78	14.52	10.37	15.14	7.05	8.20	5.73	11.47	5.02	15.57	13.30
Sheet Metal Work—Erection—Installation and Repair	2.74	4.06	1.84	5.56	1.62	4.31	2.74	4.30	2.08	2.40	1.81	4.46	2.89
Steel Erection—Doors and Sash	2.86	3.40	2.10	4.66	2.33	5.70	2.48	4.20	1.75	3.14	2.67	4.25	5.75
Steel Erection—Interior Ornament	2.86	3.40	2.10	4.66	2.33	5.70	2.48	4.20	1.75	3.14	2.67	4.25	5.75
Steel Erection—Structure	12.08	19.58	6.08	26.82	9.85	15.05	9.95	9.10	9.80	11.55	8.53	29.77	8.78
Steel Erection—Dwelling 2 Stories	8.29	7.62	4.82	10.44	5.38	9.94	8.99	4.30	5.34	7.89	3.86	15.21	7.66
Steel Erection—N.O.C.	12.08	14.16	4.87	19.40	8.61	9.85	9.95	7.50	9.80	6.89	4.87	26.91	8.78
Tile Work—Interior	1.35	1.47	1.30	2.01	1.13	2.71	2.96	1.90	1.15	1.70	1.36	2.69	2.10
Timekeepers and Watchmen	4.44	3.81	2.25	5.22	2.48	4.35	3.80	2.90	2.63	3.26	2.48	4.81	9.08
Waterproofing Interior (Brush)	3.02	5.88	1.67	8.06	3.09	7.37	2.84	4.20	1.72	2.57	2.15	3.93	6.45
Waterproofing—Trowel Interior	2.47	5.06	1.49	6.93	1.87	3.46	3.24	4.80	1.80	1.79	1.44	3.62	3.20
Waterproofing—Trowel Exterior	2.86	3.80	1.44	5.21	1.80	5.31	3.01	3.80	1.34	1.79	1.64	4.06	6.63
Waterproofing Pressure Gun	4.96	4.94	3.19	6.77	3.36	6.91	4.54	4.60	3.03	4.89	2.72	7.78	4.38
Wrecking	20.57	22.99	13.57	31.50	13.60	22.63	14.53	A	9.91	19.31	8.42	37.04	46.31

•A Specially rated—refer to company.   •B Assign governing class.   •C Subscription to State Fund Mandatory; Industrial Commission, Carson City, Nevada; Workmen's Compensation Bureau, Bismark, North Dakota; The Industrial Commission of Ohio, Columbus, Ohio;   Dept. of Labor and Industry, Olympia.

for Construction—July 1, 1969

(Base Rates in effect July 1, 1969—subject to increase or decrease according to experience ratings)

Classification of Work	Ill.	Ind.	Iowa	Kan.	Ky.	La.	Me.	Md.	Mass.	Mich.	Minn.	Miss.
Carpentry—1, 2 Family Residence	$2.98	$1.78	$1.85	$3.78	$3.40	$5.85	$2.67	$2.66	$5.36	$4.57	$3.37	$3.71
Carpentry—3 Stories or less	3.15	2.31	2.14	3.78	3.53	5.85	2.67	2.66	6.00	4.57	3.37	3.71
Carpentry—Interior Cab WK	1.56	1.15	0.91	1.67	2.30	3.26	1.20	2.10	2.66	2.94	3.37	1.99
Carpentry—General	3.93	2.78	3.02	3.83	4.83	8.87	3.62	3.68	11.82	4.57	6.31	6.74
Chimney—Construction Brick Con	11.49	7.36	6.75	13.49	13.77	26.30	8.56	12.77	12.18	17.34	16.37	13.87
Concrete Work—Bridges Culverts—C	4.66	3.43	5.03	5.55	4.94	8.21	5.94	6.91	4.59	9.52	5.03	6.28
Concrete Work Dwellings 1-2 Fam	1.56	1.75	1.40	3.46	1.99	3.89	2.45	3.15	4.03	4.44	2.20	2.04
Concrete Work—N.O.C.	5.30	2.58	2.61	4.38	3.02	6.95	3.55	4.32	9.54	5.06	5.06	4.43
Concrete or Cement Work—Floors, Sidewalks	1.85	1.18	1.47	2.51	2.75	2.87	1.45	2.88	4.03	2.53	1.85	1.99
Electrical Wiring—Inside	1.82	1.21	0.88	2.23	1.51	3.69	1.00	1.91	2.37	1.62	1.72	2.27
Excavation Earth N.O.C.	2.60	2.35	1.96	2.92	4.00	6.91	2.37	2.63	3.81	4.77	3.57	4.67
Excavation—Rock	13.29	4.95	3.88	4.86	4.64	18.78	4.69	6.11	4.32	7.02	4.44	9.54
Glaziers	3.10	2.12	2.03	2.75	3.10	7.58	3.38	4.72	5.02	3.46	4.54	5.03
Insulation Work	3.46	1.37	1.20	2.90	2.08	5.54	2.40	2.24	3.28	3.43	3.86	2.19
Lathing	1.42	1.25	1.24	2.17	2.16	4.57	1.69	2.05	3.77	1.64	1.74	2.56
Masonry	3.07	1.56	1.40	3.46	1.99	3.89	2.45	3.04	5.39	3.61	1.85	2.04
Painting and Decorating	3.15	2.04	1.81	2.31	3.48	5.19	3.09	2.90	4.15	5.49	2.66	2.51
Pile Driving	8.66	4.51	4.84	10.37	12.37	19.44	4.26	6.81	8.53	9.60	10.46	10.25
Plastering	1.81	1.37	1.20	2.95	2.08	3.58	2.40	2.24	4.31	3.43	1.85	2.19
Plumbing	1.79	1.37	1.33	2.20	1.66	5.49	1.26	1.97	2.62	1.99	2.40	2.15
Roofing	6.82	4.76	4.08	9.67	7.87	11.07	7.47	7.44	29.87	9.27	11.86	6.66
Sheet Metal Work—Erection—Installation and Repair	2.04	1.72	1.37	2.73	2.33	6.47	1.92	2.41	3.13	2.70	2.48	3.66
Steel Erection—Doors and Sash	3.14	2.53	1.59	2.73	3.88	4.80	2.34	4.70	4.23	3.64	3.25	3.45
Steel Erection—Interior Ornament	3.14	2.53	1.59	2.73	3.88	4.80	2.34	4.70	4.23	3.64	3.25	3.45
Steel Erection—Structure	14.83	8.09	7.62	12.79	11.65	12.49	7.40	15.50	14.66	16.02	11.94	6.64
Steel Erection—Dwelling 2 Stories	7.82	3.66	4.25	6.78	6.40	12.49	4.24	7.57	14.66	6.81	8.16	6.64
Steel Erection—N.O.C.	11.13	4.62	6.45	7.51	8.19	12.49	7.40	15.50	14.66	7.94	7.12	6.64
Tile Work—Interior	1.20	1.29	0.76	1.61	1.17	2.78	1.13	1.55	3.19	1.88	1.85	2.08
Timekeepers and Watchmen	3.26	2.35	2.02	3.59	3.22	6.75	2.48	3.35	7.75	3.56	3.37	3.14
Waterproofing Interior (Brush)	3.15	2.04	1.81	2.31	3.48	5.19	3.09	2.90	4.15	5.49	2.66	2.51
Waterproofing—Trowel Interior	1.81	1.37	1.20	2.95	2.08	3.58	2.40	2.24	4.31	3.43	1.85	2.19
Waterproofing—Trowel Exterior	3.07	1.56	1.40	3.46	1.99	3.89	2.45	3.04	5.39	3.61	1.85	2.04
Waterproofing Pressure Gun	5.30	2.58	2.61	4.38	3.02	6.95	3.55	4.32	9.54	5.06	5.06	4.43
Wrecking	14.06	7.99	11.41	15.73	18.40	34.70	20.09	22.70	18.06	19.81	19.97	11.27

Classification of Work	Penn.	R.I.	S.C.	S.D.	Tenn.	Tex.	Utah	Vt.	Va.	Wash. (C)	W. Va. (C)	Wisc.	Wyo. (C)
Carpentry—1, 2 Family Residence	$2.00	$3.09	$3.83	$2.00	$3.67	$5.61	$2.34	$2.16	$1.82	$3.87	$3.40	$2.05	$2.06
Carpentry—3 Stories or less	2.00	3.09	3.83	2.00	3.67	5.61	2.34	2.16	1.45	3.98	3.53	1.70	2.06
Carpentry—Interior Cab WK	2.00	1.89	1.74	1.00	1.82	3.07	2.34	1.13	1.16	1.27	2.30	2.05	2.06
Carpentry—General	2.00	4.43	5.91	2.23	4.43	7.47	2.34	2.76	2.06	5.11	4.83	3.59	2.06
Chimney—Construction Brick Con	2.30	15.93	16.33	5.76	13.81	22.82	9.08	9.42	6.27	4.20	13.77	12.76	7.99
Concrete Work—Bridges Culverts—C	2.30	2.95	6.88	3.41	6.10	8.77	3.75	4.54	3.40	5.78	4.94	3.14	3.30
Concrete Work Dwellings 1-2 Fam	2.30	2.02	1.87	1.35	2.42	5.05	2.03	1.67	1.04	3.66	1.99	2.04	1.79
Concrete Work—N.O.C.	2.30	6.49	5.02	2.51	3.90	7.16	3.16	2.65	2.84	4.20	3.02	3.46	2.78
Concrete or Cement Work—Floors, Sidewalks	1.50	2.23	2.23	1.21	2.42	3.77	1.41	1.42	1.08	1.94	2.75	2.08	1.24
Electrical Wiring—Inside	1.10	1.35	2.21	0.87	1.56	2.46	1.24	1.19	1.18	1.95	1.51	1.48	1.09
Excavation Earth N.O.C.	2.70	2.77	3.21	2.24	2.85	5.52	2.13	4.41	2.58	4.40	4.00	2.79	2.49
Excavation—Rock	2.70	7.82	8.51	3.45	4.25	7.45	3.33	4.08	2.46	9.32	4.64	6.95	2.93
Glaziers	1.35	4.33	4.47	2.44	4.37	2.78	1.24	3.39	2.15	4.35	3.10	2.73	1.09
Insulation Work	2.00	2.26	2.67	1.57	2.09	5.32	2.24	1.93	1.34	3.40	2.18	1.73	1.97
Lathing	1.20	2.52	2.58	1.19	2.54	1.82	1.68	1.55	1.19	3.30	2.16	1.35	1.48
Masonry	2.25	2.02	1.87	1.35	2.42	5.05	1.82	1.67	1.10	3.77	1.99	1.70	1.60
Painting and Decorating	2.70	3.11	3.92	1.69	2.61	3.66	1.32	2.38	1.95	4.19	3.48	2.04	1.16
Pile Driving	3.80	6.05	8.13	6.49	10.13	14.62	4.50	4.45	4.50	9.79	12.37	6.87	3.96
Plastering	1.20	2.26	2.67	1.57	2.09	3.84	2.24	1.93	1.34	3.40	2.08	1.33	1.97
Plumbing	1.35	1.99	1.93	0.94	1.67	3.44	1.78	1.30	1.16	1.72	1.66	1.85	1.57
Roofing	3.50	5.15	5.93	5.13	5.98	11.07	6.06	8.13	4.81	3.75	7.87	7.41	5.33
Sheet Metal Work—Erection—Installation and Repair	3.50	2.34	3.43	1.74	2.85	4.36	1.59	1.79	1.59	3.75	2.33	1.96	1.38
Steel Erection—Doors and Sash	5.90	3.03	2.29	1.71	2.48	4.56	1.57	2.18	1.65	10.56	3.88	2.36	1.38
Steel Erection—Interior Ornament	5.90	3.03	2.29	1.71	2.48	4.56	1.57	2.18	1.65	10.56	3.88	2.36	1.38
Steel Erection—Structure	5.90	19.34	11.38	7.60	6.42	16.88	6.22	12.13	6.99	15.86	11.65	6.72	5.47
Steel Erection—Dwelling 2 Stories	5.90	8.26	6.88	3.33	5.66	16.88	5.18	5.97	3.58	7.51	6.40	4.51	4.56
Steel Erection—N.O.C.	5.90	19.34	7.95	3.35	6.42	16.88	6.22	7.82	5.13	10.56	8.19	8.16	5.47
Tile Work—Interior	1.70	1.74	1.38	0.87	1.82	2.30	1.44	1.07	1.05	1.38	1.17	1.58	1.27
Timekeepers and Watchmen	B	3.76	3.22	1.97	3.77	2.14	2.13	2.33	1.91	4.18	3.22	2.78	2.14
Waterproofing Interior (Brush)	2.70	3.11	3.92	1.69	2.61	3.66	1.32	2.38	1.95	4.19	3.48	2.04	1.16
Waterproofing—Trowel Interior	1.20	2.26	2.67	1.57	2.09	3.84	2.24	1.93	1.34	3.40	2.08	1.33	1.97
Waterproofing—Trowel Exterior	2.25	2.02	1.87	1.35	2.42	5.05	1.82	1.67	1.10	3.77	1.99	1.70	1.60
Waterproofing Pressure Gun	2.30	6.49	5.02	2.51	3.90	7.16	3.16	2.65	2.84	4.20	3.02	3.46	2.78
Wrecking	15.05	15.62	20.61	10.36	19.11	31.72	13.21	13.70	7.59	21.52	18.40	8.04	11.62

Washington; State Compensation Commission, Charleston, W. Virginia; Workmen's Compensation, State Treasurer's Office, Cheyenne, Wyoming.

N.O.C. – Not otherwise covered.

Appendix D

PUBLIC LIABILITY AND PROPERTY DAMAGE INSURANCE

Typical insurance classifications are listed here together with their corresponding liability code numbers.

The rates given here are characteristic basic manual rates per $100 payroll applicable to each classification. As in the case of Workmen's Compensation rating, there are a number of limitations or qualifications regarding division of payroll on a given job or location between certain insurance classifications. Accordingly, it is recommended that these classifications and rates be used only as a general guide. This information is being reprinted from *The Building Estimator's Reference Book* by permission of the Frank Walker Company.

Public Liability and Property Damage Insurance Classifications

Description	Code Number
Boiler installation or repair—steam—or tank erection or repair within buildings	3436
Carpentry—1 and 2 family residences	5645
Carpentry—dwellings, 3 stories or less	5645
Carpentry—cabinets and interior trim	5437
Carpentry—general	3457
Cofferdam work—not pneumatic	3438
Concrete construction—bridges	5213
Concrete construction—general	5213
Concrete or cement floors, walks, etc.:	
Floors	5213
Sidewalks, driveways	5200
Concrete work incidental to private residences	5213
Door, door frame or sash erection—metal or metal covered	3442
Electrical apparatus installation or repair	3436
Electrical wiring—within buildings, including fixtures and appliances	5190
Excavation—earth—and grading of land:	
General	3470
Grading of land	6041
Excavation—rock	3470
Fence construction—metal	3442
Furniture or fixtures installation in offices or stores—portable—metal or wood	5146
Glazing	5437
Insulation work—buildings (1)	5480
Iron, brass or bronze work—non-structural—within buildings	3442
Iron or steel erection—metal bridges, frame structures, iron work on the outside of buildings:	
Bridges	5067
Other	3452
Iron or steel erection—frame structures not over two stories	3452
Iron or steel erection—dwellings not over 2 stories	3442
Iron or steel erection—general	5057
Lathing	5443
Masonry	3447
Painting, decorating or paper hanging	3429

Public Liability and Property Damage Insurance Classifications (Cont.)

Description	Code Number
Painting—metal bridges or metal structures over two stories in height (Painting ship hulls or oil or gasoline tanks to be separately rated)	5067
Pile driving:	
Building foundations	3470
General	3430
Plastering	5480
Plumbing	3434
Roofing	5551
Sheet metal erection	5538
Steam pipe or boiler insulation	3434
Tile, stone, mosaic or terrazzo work—interior	3456
Timekeepers and watchmen (2)	5610
Waterproofing (3)	:::
Wrecking—general—not marine	3451

Notes: (1) Class available only when insulation or accoustical work is performed as a separate operation, not a part of, or incidental to any other construction operations performed by the same contractor at the same job or location.

(2) This classification is not applicable to the payroll for watchmen or timekeepers except when the payroll for watchmen, timekeepers and cleaners is more than all other payroll of the insured which is subject to contruction or erection classifications at the same job or location.

(3) Waterproofing, other than roofing or subaqueous work, when performed as a separate operation not a part of other operations by the same contractor:

  (a) Applied by brush—see Painting 3429

  (b) Applied by trowel—interior—see Plastering 5480; exterior—see Plastering 3447

  (c) Applied by spray or pressure gun—see Concrete Construction—general 5213.

Footnotes for Following P.L. Rate Tables

(a) Submit for individual rating.

s Table A applies for increasing limits above standard limits.

x Rate excludes coverage for damage arising out of blasting or explosion.

c Rate excludes coverage for damage arising out of collapse of or structural damage to any building or structure due to grading, excavation, pile driving, underpinning, demolition operations, etc.

u Rate excludes coverage for damage to underground wires, sewers, mains, etc., caused by use of mechanical equipment for grading, paving, excavating or drilling.

(The exclusions referred to above are defined in greater detail in the Liability Manual and usually can be removed for an additional premium charge.)

** This rate contemplates $50.00 deductible per claim for damage arising out of spray painting operations.

Manual Rates for Public Liability and Property Damage Insurance

Code	Ala.	Alas.	Ariz.	Ark.	Calif.	Colo.	Conn.	Dela.	D.C.	Fla.	Ga.
1-3436	.14	.08	.12	.17	.22	.09	.23	.06	.26	.16	.15
2-5645	.11	.08	.10	.10	.17	.06	.37	.10	.32	.11	.11
3-5645	.11	.08	.10	.10	.17	.06	.37	.10	.32	.11	.11
4-5437	.15	.11	.13	.14	.18	.06	.27	.08	.14	.15	.11
5-3457	.21	.15	.20	.23	.40	.13	.52	.12	.40	.16	.16
6-3438	.43	.34	.34	.45	.26	.14	.37	.11	.22	.23	.16
7-5213	.24	.18	.21	.26	.29	.18	.42	.11	.25	.25	.26
8-5213	.24	.18	.21	.26	.29	.18	.42	.11	.25	.25	.26
9-5213	.24	.18	.21	.26	.29	.18	.42	.11	.25	.25	.26
10-5200	.24	.18	.21	.26	.29	.18	.42	.11	.25	.26	.26
11-5213	.24	.18	.21	.26	.29	.18	.42	.11	.25	.26	.20
12-3442	.19	.22	.23	.20	.28	.18	.37	.13	.25	.16	.15
13-3436	.14	.08	.12	.17	.22	.06	.23	.06	.26	.11	.10
14-5190	.10	.05	.07	.12	.12	.09	.09	.045	.26	.11	.63
15-3470	.86	.77	.78	.89	.94	.42	1.03	.36	.99	.86	.63
16-6041	.17	.16	.21	.18	.28	.12	.42	.14	.16	.22	.22
17-3470	.86	.77	.78	.89	.94	.42	1.03	.36	.99	.86	.63
18-3442	.19	.22	.23	.20	.28	.18	.37	.13	.25	.26	.15
19-5146	.07	.07	.08	.06	.12	.07	.14	.046	.12	.09	.08
20-5437	.15	.15	.13	.14	.18	.10	.27	.08	.14	.15	.11
21-5480	.11	.07	.07	.12	.21*	.05	.14	.033	.08	.08	.07
22-3442	.19	.22	.23	.20	.28	.18	.37	.13	.25	.26	.15
23-5067	.43	.47	.52	.44	.77	.41	.72	.31	.56	.60	.44
24-3452	.78	.84	.87	.82	1.29	.66	1.75	.52	1.10	.94	.72
25-3452	.78	.84	.87	.82	1.29	.66	1.75	.52	1.10	.94	.72
26-3442	.19	.22	.23	.20	.28	.18	.37	.13	.25	.26	.15
27-5057	.53	.49	.60	.55	.94	.52	.84	.45	.71	.74	.66
28-5443	.06	.09	.10	.06	.09	.14	.14	.05	.10	.08	.06
29-3447	.14	.12	.15	.17	.21	.11	.32	.07	.20	.15	.12
30-3429	.043	.07	.09	.048	.11	.039	.14	.036	.08	.08	.09
31-5067	.43	.47	.52	.44	.77	.41	.72	.31	.56	.60	.44
32-3470	.86	.77	.78	.89	.94	.42	1.03	.36	.99	.86	.63
33-3430	.07	.09	.10	.08	.19	.09	.19	.036	.10	.18	.10
34-5480	.11	.07	.07	.12	.21*	.05	.14	.033	.08	.08	.07
35-3434	.17	.18	.20	.20	.19	.11	.22	.08	.22	.08	.09
36-5551	.21	.18	.21	.21	.34	.15	.53	.13	.31	.14	.30
37-5538	.10	.07	.08	.11	.12	.06	.14	.048	.09	.23	.06
38-3434	.17	.18	.20	.20	.19	.11	.22	.08	.22	.10	.09
39-3456	.06	.09	.10	.06	.09	.06	.14	.05	.10	.14	.06
40-5610	.46	.49	.54	.46	.83	.42	.98	.33	.56	.62	.46
41-3451s	.97	1.06	1.12	1.18	1.73	.84	2.35	.69	1.40	1.70	1.28

*Calif.#3750

Manual Rates for Public Liability and Property Damage Insurance

Code	H.I.	Ida.	Ill.	Ind.	Ia.	Kan.	Ky.	La.	Me.	Md.	Mass.
1-3436	.11	.08	.17	.08	.11	.05	.10	.18	.05	.09	.16
2-5645	.09	.08	.22	.09	.13	.07	.09	.14	.08	.12	.17
3-5645	.09	.08	.22	.09	.13	.07	.09	.14	.08	.12	.17
4-5437	.09	.11	.24	.08	.11	.07	.09	.18	.07	.11	.15
5-3457	.13	.15	.43	.14	.17	.10	.11	.25	.10	.13	.46
6-3438	.11	.34	.32	.10	.18	.11	.12	.18	.09	.17	.20
7-5213	.13	.18	.37	.13	.18	.12	.16	.25	.09	.11	.28
8-5213	.13	.18	.37	.13	.19	.12	.16	.25	.09	.11	.28
9-5213	.13	.18	.37	.13	.19	.12	.16	.25	.09	.11	.28
10-5200	.13	.18	.37	.13	.19	.12	.16	.25	.09	.11	.28
11-5213	.13	.18	.37	.13	.19	.12	.16	.25	.09	.11	.28
12-3442	.18	.22	.39	.11	.19	.12	.15	.41	.11	.17	.25
13-3436	.11	.08	.17	.08	.11	.05	.10	.18	.05	.09	.16
14-5190	.07	.05	.14	.05	.08	.04	.08	.14	.038	.06	.08
15-3470	.33	.77	.76	.28	.58	.34	.51	.49	.30	.39	.67
16-6041	.15	.16	.30	.18	.18	.11	.14	.19	.12	.13	.25
17-3470	.33	.77	.76	.28	.58	.34	.51	.49	.30	.39	.67
18-3442	.18	.22	.39	.11	.22	.12	.15	.41	.11	.17	.25
19-5146	.05	.07	.16	.08	.07	.036	.06	.08	.039	.05	.08
20-5437	.09	.11	.24	.08	.11	.07	.09	.18	.07	.11	.15
21-5480	.05	.07	.11	.08	.07	.038	.048	.08	.028	.06	.09
22-3442	.18	.22	.39	.11	.22	.12	.15	.41	.11	.17	.25
23-5067	.33	.47	.73	.28	.46	.29	.34	.60	.26	.44	.52
24-3452	.59	.84	1.29	.49	.84	.49	.56	1.16	.44	.78	.77
25-3452	.59	.84	1.29	.49	.84	.49	.56	1.16	.44	.78	.77
26-3442	.18	.22	.39	.11	.11	.12	.15	.41	.11	.17	.25
27-5057	.39	.49	.79	.31	.57	.33	.39	.74	.38	.49	.58
28-5443	.044	.09	.11	.07	.06	.035	.048	.13	.042	.044	.07
29-3447	.11	.12	.23	.09	.16	.09	.13	.19	.06	.09	.17
30-3429	.05	.07	.10	.05	.07	.041	.06	.11	.03	.07	.07
31-5067	.33	.47	.73	.28	.28	.46	.34	.60	.26	.44	.52
32-3470	.33	.77	.76	.049	.58	.34	.51	.49	.30	.39	.67
33-3430	.07	.09	.16	.08	.09	.06	.07	.11	.05	.12	.17
34-5480	.05	.07	.11	.07	.07	.038	.048	.08	.028	.06	.09
35-3434	.10	.18	.17	.07	.14	.07	.09	.19	.07	.09	.11
36-555	.17	.18	.38	.18	.25	.18	.18	.54	.11	.20	.59
37-5538	.06	.07	.11	.07	.10	.06	.09	.08	.04	.08	.09
38-3434	.10	.18	.17	.07	.14	.07	.09	.19	.07	.09	.11
39-3456	.044	.09	.11	.07	.06	.035	.048	.13	.042	.044	.07
40-5610	.36	.49	.90	.30	.48	.31	.36	.57	.28	.45	.43
41 3451s	.71	1.06	1.74	.59	.89	.56	1.19	1.09	.58	1.04	1.08

Manual Rates for Public Liability and Property Damage Insurance

Code	Mich.	Minn.	Miss.	Mo.	Mont.	Nebr.	Nev.	N.H.	N.J.	N.M.	N.Y.-1
1-3436	.07	.13	.18	.15	.08	.08	.12	.05	.22	.12	1.15
2-5645	.08	.26	.13	.16	.08	.10	.10	.08	.23	.10	.79
3-5645	.08	.26	.13	.16	.08	.10	.10	.08	.23	.10	.79
4-5437	.07	.13	.18	.19	.11	.10	.13	.07	.22	.13	.93
5-3457	.12	.21	.32	.36	.15	.12	.20	.10	.41	.20	2.72
6-3438	.13	.20	.56	.21	.34	.16	.34	.09	.16	.34	1.22
7-5213	.14	.23	.30	.32	.18	.16	.21	.09	.30	.21	2.41
8-5213	.14	.23	.30	.32	.18	.16	.21	.09	.30	.21	2.41
9-5213	.14	.23	.30	.32	.18	.16	.21	.09	.30	.21	2.41
10-5200	.12	.23	.30	.32	.18	.16	.21	.09	.30	.21	2.41
11-5213	.14	.23	.30	.32	.18	.16	.21	.09	.30	.21	2.41
12-3442	.12	.22	.24	.29	.22	.17	.23	.11	.39	.23	1.67
13-3436	.07	.13	.18	.15	.08	.12	.12	.05	.22	.12	1.15
14-5190	.06	.07	.13	.13	.05	.06	.07	.038	.10	.07	.80
15-3470	.35	.85	1.16	.82	.77	.44	.78	.30	.70	.78	4.23
16-6041	.08	.14	.22	.25	.16	.13	.21	.12	.22	.21	1.13
17-3470	.35	.85	1.16	.82	.77	.44	.78	.30	.70	.78	4.23
18-3442	.12	.22	.24	.29	.22	.17	.23	.11	.39	.23	1.67
19-5146	.036	.10	.08	.12	.07	.06	.08	.039	.14	.08	.62
20-5437	.07	.13	.18	.19	.11	.10	.13	.07	.22	.13	.93
21-5480	.042	.08	.13	.10	.07	.07	.07	.028	.10	.07	.59
22-3442	.12	.22	.24	.29	.22	.17	.23	.11	.39	.23	1.67
23-5067	.30	.52	.55	.62	.47	.40	.52	.26	.58	.52	4.54
24-3452	.53	.81	.98	1.21	.84	.69	.87	.44	1.16	.87	9.49
25-3452	.53	.81	.98	1.21	.84	.69	.87	.44	1.16	.87	9.49
26-3442	.12	.22	.24	.29	.22	.17	.23	.11	.39	.23	1.67
27-5057	.31	.70	.67	.76	.49	.47	.60	.38	.78	.60	3.98
28-5443	.031	.08	.07	.09	.07	.05	.08	.042	.11	.10	.74
29-3447	.08	.14	.16	.21	.12	.12	.15	.06	.31	.15	2.40
30-3429	.038	.05	.05	.08	.07	.05	.09	.03	.14	.09	.57
31-5067	.30	.52	.55	.62	.47	.40	.52	.26	.58	.52	4.54
32-3470	.35	.85	1.16	.82	.77	.44	.78	.30	.70	.78	4.23
33-3430	.05	.10	.10	.13	.09	.08	.10	.05	.16	.10	.86
34-5480	.042	.08	.13	.10	.07	.07	.07	.028	.10	.07	.59
35-3434	.10	.17	.32	.22	.18	.20	.20	.07	.16	.20	1.09
36-5551	.12	.23	.25	.25	.18	.15	.21	.11	.34	.21	1.38
37-5538	.039	.09	.13	.14	.07	.08	.08	.04	.13	.08	.80
38-3434	.10	.17	.32	.22	.18	.20	.20	.07	.16	.20	1.09
39-3456	.031	.08	.07	.09	.09	.05	.10	.042	.11	.10	.74
40-5610	.31	.54	.57	.66		.16	.54	.28	.54	.54	6.67
41-3451s	.61	1.17	1.66	1.83	1.06	.77	1.12	.58	1.31	1.12	4.43

Manual Rates for Public Liability and Property Damage Insurance

Code	N.Y.-2	N.C.	N.D.	Ohio	Okla.	Ore.	Pa.	R.I.	S.C.	S.D.	Tenn.
1-3436	.34	.09	.09	.11	.18	.11	.17	.17	.16	.09	.15
2-5645	.26	.06	.06	.13	.13	.10	.16	.24	.09	.06	.12
3-5645	.26	.06	.06	.13	.13	.10	.16	.24	.09	.06	.12
4-5437	.30	.07	.10	.12	.18	.12	.14	.20	.11	.10	.16
5-3457	.61	.11	.13	.20	.26	.17	.19	.32	.15	.13	.18
6-3438	.35	.10	.14	.18	.25	.18	.21	.19	.14	.14	.21
7-5213	.65	.20	.18	.21	.28	.21	.25	.27	.31	.18	.28
8-5213	.65	.20	.18	.21	.28	.21	.25	.27	.31	.18	.28
9-5213	.65	.20	.18	.21	.28	.21	.25	.27	.31	.18	.28
10-5200	.65	.20	.18	.21	.28	.21	.25	.27	.31	.18	.28
11-5213	.65	.20	.18	.21	.28	.21	.25	.27	.31	.18	.28
12-3442	.52	.12	.18	.19	.36	.21	.19	.31	.19	.18	.26
13-3436	.34	.09	.09	.11	.18	.11	.17	.17	.16	.09	.15
14-5190	.28	.07	.06	.08	.09	.06	.11	.12	.11	.06	.12
15-3470	1.69	.40	.42	.61	.94	.83	.66	.88	.52	.42	.95
16-6041	.43	.19	.12	.23	.20	.23	.11	.21	.30	.12	.30
17-3470	1.69	.40	.42	.61	.94	.83	.66	.88	.52	.42	.95
18-3442	.52	.12	.18	.19	.36	.21	.19	.31	.19	.18	.26
19-5146	.24	.048	.07	.07	.11	.07	.07	.11	.08	.07	.09
20-5437	.30	.07	.10	.12	.18	.12	.14	.20	.11	.10	.16
21-5480	.17	.06	.05	.07	.11	.07	.07	.09	.10	.05	.09
22-3442	.52	.12	.18	.19	.36	.21	.19	.31	.19	.18	.26
23-5067	1.10	.52	.41	.49	.66	.49	.61	.72	.80	.41	.59
24-3452	2.66	.42	.66	.75	2.97	.81	.74	1.32	.64	.66	.83
25-3452	2.66	.42	.66	.75	2.97	.81	.74	1.32	.64	.66	.83
26-3442	.52	.12	.18	.19	.36	.21	.19	.31	.19	.18	.26
27-5057	1.39	.34	.52	.54	.84	.56	.57	.82	.51	.52	.73
28-5443	.20	.039	.06	.07	.07	.07	.10	.09	.06	.06	.09
29-3447	.64	.09	.11	.10	.22	.15	.12	.20	.13	.11	.19
30-3429	.17	.031	.039	.08	.08	.08	.08	.11	.043	.039	.06
31-5067	1.10	.52	.41	.49	.66	.49	.61	.72	.80	.41	.59
32-3470	1.69	.40	.42	.61	.94	.83	.66	.88	.52	.42	.95
33-3430	.29	.06	.09	.09	.13	.10	.09	.14	.09	.09	.12
34-5480	.17	.06	.05	.07	.11	.07	.07	.09	.10	.05	.09
35-3434	.36	.09	.11	.13	.26	.16	.14	.15	.14	.11	.13
36-5551	.65	.16	.15	.19	.47	.22	.23	.31	.23	.15	.22
37-5538	.22	.06	.06	.08	.11	.11	.12	.09	.08	.06	.08
38-3434	.36	.09	.11	.13	.26	.16	.14	.15	.14	.11	.13
39-3456	.20	.039	.06	.07	.09	.08	.10	.09	.06	.06	.09
40-5610	2.27	.29	.42	.53	.67	.51	.40	.75	.45	.42	.62
41-3451s	1.85	.89	.84	1.49	1.65	1.13	1.18	1.51	1.43	.84	1.64

ESTIMATOR'S REFERENCE BOOK

Manual Rates for Public Liability and Property Damage Insurance

Code	Tex.	Utah	Vt.	Va.	Wash.	W. Va.	Wis.	Wyo.
1-3436	.14	.12	.05	.10	.11	.06	.16	.09
2-5645	.08	.10	.08	.12	.11	.10	.17	.06
3-5645	.08	.10	.08	.12	.11	.10	.17	.06
4-5437	.16	.13	.07	.10	.12	.08	.17	.10
5-3457	.26	.20	.10	.16	.17	.12	.20	.13
6-3438	.28	.34	.09	.13	.18	.11	.20	.14
7-5213	.22	.21	.09	.16	.23	.11	.24	.18
8-5213	.22	.21	.09	.16	.23	.11	.24	.18
9-5213	.22	.21	.09	.16	.23	.11	.24	.18
10-5200	.22	.21	.09	.16	.23	.11	.24	.18
11-5213	.22	.21	.09	.16	.22	.13	.24	.18
12-3442	.33	.23	.11	.16	.22	.13	.27	.18
13-3436	.14	.12	.05	.10	.11	.06	.16	.09
14-5190	.13	.07	.038	.06	.06	.045	.11	.06
15-3470	.54	.78	.30	.47	.74	.36	.50	.42
16-6041	.18	.21	.12	.17	.20	.14	.17	.12
17-3470	.54	.78	.30	.47	.74	.36	.50	.42
18-3442	.33	.23	.11	.16	.22	.13	.27	.18
19-5146	.13	.08	.039	.07	.07	.046	.07	.07
20-5437	.16	.13	.07	.10	.12	.08	.17	.10
21-5480	.11	.07	.028	.05	.07	.033	.09	.05
22-3442	.33	.23	.11	.16	.22	.13	.27	.18
23-5067	.87	.52	.26	.38	.49	.31	.54	.41
24-3452	1.18	.87	.44	.76	.83	.52	1.13	.66
25-3452	1.18	.87	.44	.76	.83	.52	1.13	.66
26-3442	.33	.23	.11	.16	.22	.13	.27	.18
27-5057	.89	.60	.38	.62	.57	.45	.57	.52
28-5443	.08	.10	.042	.07	.08	.05	.08	.06
29-3447	.17	.15	.06	.08	.13	.07	.14	.11
30-3429	.08	.09	.03	.05	.09	.036	.07	.039
31-5067	.87	.52	.26	.38	.49	.31	.54	.41
32-3470	.54	.78	.30	.47	.74	.36	.50	.42
33-3430	.13	.10	.05	.08	.10	.06	.10	.09
34-5480	.11	.07	.028	.05	.07	.033	.09	.05
35-3434	.15	.20	.07	.12	.15	.08	.16	.11
36-5551	.26	.21	.11	.16	.21	.13	.25	.15
37-5538	.11	.08	.04	.06	.07	.048	.07	.06
38-3434	.15	.20	.07	.12	.15	.08	.16	.11
39-3456	.08	.10	.042	.07	.08	.05	.08	.06
40-5610	.77	.54	.28	.38	.52	.33	.55	.42
41-3451s	1.42	1.12	.58	.86	1.07	.69	1.18	.84

Manual Rates for Public Liability and Property Damage Insurance

Code	N.Y. Terr 1 Property Damage	N.Y. Terr 2 P.D.	State Group 1 P.D.	State Group 2 P.D.	
1-3436	.41	.26	.21	.19	*New York Terr. #1*
2-5645	.14	.14	.13	.12	All of New York City
3-5645	.14	.14	.13	.12	except Burrough of
4-5437	.25	.25	.37	.35	Richmond and
5-3457	.54	.23	.18	.17	Governor's Island
6-3438	(a)xcu	(a)xcu	(a)xcu	(a)xcu	
7-5213	.35	.24	.16	.15	*New York Terr. #2*
8-5213	.35	.24	.16	.15	
9-5213	.35	.24	.16	.15	
10-5200	.28	.28	.19	.18	Remainder of New York
11-5213	.35	.24	.16	.15	
12-3442	.28	.28	.31	.29	
13-3436	.41	.26	.21	.19	*State Group #1*
14-5190	.20	.20	.19	.16	
15-3470	1.61xcu	1.61xcu	1.02xcu	.93xcu	District of Columbia
16-6041	.77xcu	.77xcu	.60xcu	.60xcu	Florida
17-3470	1.61xcu	1.61xcu	1.02xcu	.93xcu	Georgia
18-3442	.28	.28	.31	.29	Iowa
19-5146	.28	.28	.18	.17	Louisiana
20-5437	.25	.25	.37	.35	Missouri
21-5480	.18	.18	.23	.22*	Oklahoma
22-3442	.28	.28	.31	.29	Texas
23-5067	6.00	1.07	1.32	1.28	Wisconsin
24-3452	1.19	.91	.62	.57	
25-3452	1.19	.91	.62	.57	
26-3442	.28	.28	.31	.29	
27-5057	.97	.72	.56	.51	*State Group #2*
28-5443	.13	.13	.09	.08	
29-3447	.31	.19	.12	.11	Remainder of country
30-3429	.55	.40	.43**	.39**	
31-5067	6.00	1.07	1.32**	1.28**	
32-3470	1.61xcu	1.61xcu	1.02xcu	.93xcu	
33-3430	.88xcu	.88xcu	.60xcu	.58xcu	
34-5480	.18	.18	.23	.22*	
35-3434	.55u	.55u	.42u	.36u	
36-5551	1.10	1.10	.95	.90	
37-5538	.35	.28	.27	.25	
38-3434	.55	.55	.42	.36	
39-3456	.25	.25	.36	.34	
40-5610	.54	.54	.23	.23	
41-3451s	(a)xc	(a)xc	(a)xc	(a)xc	

*Calif. 16¢

For Footnotes and Explanation of Symbols See Page 35

Appendix E

CONTRACTORS' EQUIPMENT OWNERSHIP EXPENSE

The percentage rates set up for the various items represent an approximate average of conditions under which the equipment has been operated. These rates are subject to adjustment to fit the experience of the individual contractor.

These schedules are designed to reflect ownership cost and do not contain any profit for the equipment.

These schedules have gained wide usage and acceptance as industry practice, not only among contractors but in the professions as well. All interested in the business condition of the contractor in connection with contracts or credit have found use for the schedule.

However, the author of these schedules (the Associated General Contractors of America) cautions as follows, "In all such applications, attention is directed to the intended operation of the schedule. The rates are not determined by any precise method and should always be applied in the light of individual experience."

These schedules do not carry equipment prices. Each contractor must apply these percentages to the cost of his own equipment.

The figure shown in the "Total Ownership Expenses" column is the cost of operation for a year expressed in terms of percentage of the cost of the particular piece of equipment in question.

To get the ownership cost per month, which is shown in the column entitled "Expense Per Working Month Percent," the "Total Ownership Expense" must be divided by the "Average Use in Months per Year." As an example, if a piece of equipment was purchased for $36,000 and the total ownership cost was 30% and average use per year was 6 months, then the cost per month in percent would be 30 (%) divided by 6 (months) = 5%. And the actual ownership cost per month would be 5% of $36,000 = $1,800.

In computing working time it is usually customary to consider a month to be composed of 22 working days or 176 hours.

EQUIPMENT	AVERAGE ANNUAL EXPENSE PERCENT OF CAPITAL INVESTMENT WITHOUT FIELD REPAIRS				AVERAGE USE MONTHS PER YEAR	EXPENSE PER WORKING MONTH PERCENT	APPLICATION OF A.G.C. SCHEDULE TO OWNER'S VALUES	
	Depreciation	Overhauling, Major Repairs Painting	Interest Taxes Storage Insurance	Total Ownership Expense			Value Dollars	Expense Per Working Month Dollars
							(Fill in your own values)	

AIR COMPRESSOR, PORTABLE

High pressure—Free air delivered at 100 p.s.i. two stage and single stage, water or air cooled

Gasoline

20-185 cfm.	25	15	11	51	6	8.5	_____	_____
210-900 cfm.	20	15	11	46	6	7.7	_____	_____

Diesel

86-185 cfm.	25	15	11	51	6	8.5	_____	_____
210-1200 cfm.	20	15	11	46	6	7.7	_____	_____

AIR TOOLS AND ACCESSORIES

Hammers

Calking and chipping	33	10	11	54	6	9.0	_____	_____
Pavement breaking, 20–90 lb.	33	10	11	54	6	9.0	_____	_____
Riveting, 1/2 in. rivet, 14 lbs.	33	10	11	54	6	9.0	_____	_____
Riveting, 1 in. rivet, 25 lbs.	33	10	11	54	6	9.0	_____	_____
Rivet buckerup	33	10	11	54	6	9.0	_____	_____
Rivet hammer, rivet buster	33	10	11	54	6	9.0	_____	_____
Sheeting driver, 120 lbs.	33	10	11	54	6	9.0	_____	_____

Hoists

500 lbs., single drum	17	10	11	38	8	4.8	_____	_____
750–5000 lbs., single drum	13	10	11	34	8	4.3	_____	_____
1800–2400 lbs., double drum	13	10	11	34	8	4.3	_____	_____
2400–5000 lbs., double drum	17	13	11	41	8	5.1	_____	_____

Hose

5/8 in.–2 in. diam.	50	10	11	71	8	8.9	_____	_____
Whip	50	10	11	71	8	8.9	_____	_____

Jackhammer

Small, medium or large	33	10	11	54	6	9.0	_____	_____

EQUIPMENT	AVERAGE ANNUAL EXPENSE PERCENT OF CAPITAL INVESTMENT WITHOUT FIELD REPAIRS				AVERAGE USE MONTHS PER YEAR	EXPENSE PER WORKING MONTH PERCENT	APPLICATION OF A.G.C. SCHEDULE TO OWNER'S VALUES	
	Depreciation	Overhauling, Major Repairs Painting	Interest Taxes Storage Insurance	Total Ownership Expense			Value Dollars	Expense Per Working Month Dollars
							(Fill in your own values)	

AIR TOOLS AND ACCESSORIES (Cont'd.)
Tampers

Backfill, all sizes	33	10	11	54	6	9.0	_____	_____

Wrenches, (without sockets)
Standard type; 1/4 in.—4 in. bolt size

capacity..	25	10	11	46	6	7.7	_____	_____
Controlled torque, 3/8 in.—7/8 in.	25	10	11	46	6	7.7	_____	_____

COMPACTION EQUIPMENT
Compactors
Rammer type, manually guided, gasoline or

electric, all sizes	25	15	11	51	4	12.8	_____	_____
Vibratory plate type, manually guided, gasoline or electric, all sizes......................	29	18	11	58	5	11.6	_____	_____
Vibratory plate type, self-propelled, all sizes.	29	15	11	55	6	9.2	_____	_____

CONCRETE PRODUCING EQUIPMENT

Buggies, concrete
Power operated, pneumatic tired

Operator walking type, 6–12 cu. ft.	28	17	11	56	6	9.3	_____	_____
Operator riding type, to 18 cu. ft.	28	20	11	59	6	9.8	_____	_____

Buggies, mortar

Manually operated, cushion tired, 6 cu. ft.	29	13	11	53	6	8.8	_____	_____

CONVEYOR, Elevating Belt—See also Loader
Portable, no power, 18–32 in. width—all

lengths	25	18	11	54	6	9.0	_____	_____
Portable, with power, 18–24 in. width—all lengths	32	20	11	63	6	10.5	_____	_____

EQUIPMENT	AVERAGE ANNUAL EXPENSE PERCENT OF CAPITAL INVESTMENT WITHOUT FIELD REPAIRS				AVERAGE USE MONTHS PER YEAR	EXPENSE PER WORKING MONTH PERCENT	APPLICATION OF A.G.C. SCHEDULE TO OWNER'S VALUES	
	Deprecia-tion	Over-hauling, Major Repairs Painting	Interest Taxes Storage Insurance	Total Owner-ship Expense			Value Dollars	Expense Per Working Month Dollars
							(Fill in your own values)	

COMPACTION EQUIPMENT (Cont'd.)

EQUIPMENT								
30–36 in. width—all lengths	32	20	11	63	6	10.5	_____	_____
42–48 in. width—all lengths	32	20	11	63	6	10.5	_____	_____
60–72 in. width (Top Loading) —all lengths..	25	20	11	56	6	9.3	_____	_____

CRANES
Crawler

EQUIPMENT								
Diesel, 4 tons, 10 ft. radius	25	10	11	46	8	5.8	_____	_____
6—12 tons, 10 ft. radius	20	10	11	41	8	5.1	_____	_____
7—12 tons, 12 ft. radius	20	10	11	41	8	5.1	_____	_____
9—12 tons, 45 ft. radius	20	10	11	41	8	5.1	_____	_____
13—24-1/2 tons, 45 ft. radius	17	10	11	38	7	5.4	_____	_____
14—18 tons, 12 ft. radius.................	17	10	11	38	7	5.4	_____	_____
18 tons, 18 ft. radius....................	17	10	11	38	7	5.4	_____	_____
18 tons, 20 ft. radius....................	17	10	11	38	7	5.4	_____	_____
20—50 tons, 12 ft. radius.................	13	10	11	34	7	4.9	_____	_____
25 tons, 14 ft. radius.................	13	10	11	34	7	4.9	_____	_____
25 tons, 20 ft. radius.................	13	10	11	34	7	4.9	_____	_____
50—100 tons, 12 ft. radius.................	13	10	11	34	6	5.7	_____	_____
60—72-1/2 tons, 14 ft. radius	14	10	11	35	6	5.8	_____	_____
100—125 tons, 14 ft. radius..............	13	10	11	34	6	5.7	_____	_____
to 200 tons, 14 ft. radius.................	13	10	11	34	6	5.7	_____	_____
Gasoline, 4—7 tons, 10 ft. radius.................	25	10	11	46	8	5.8	_____	_____
7 tons, 12 ft. radius	25	10	11	46	8	5.8	_____	_____
8 tons, 10 ft. radius	20	10	11	41	8	5.1	_____	_____
8—10 tons, 12 ft. radius	20	10	11	41	8	5.1	_____	_____
9—12 tons, 45 ft. radius..............	20	10	11	41	8	5.1	_____	_____
13—24-1/2 tons, 45 ft. radius	17	10	11	38	7	5.4	_____	_____
14—15 tons, 12 ft. radius.............	17	10	11	38	7	5.4	_____	_____
15 tons, 20 ft. radius.................	17	10	11	38	7	5.4	_____	_____
18—40 tons, 12 ft. radius.............	13	10	11	34	7	4.9	_____	_____
25 tons, 20 ft. radius.................	13	10	11	34	7	4.9	_____	_____

EQUIPMENT	AVERAGE ANNUAL EXPENSE PERCENT OF CAPITAL INVESTMENT WITHOUT FIELD REPAIRS				AVERAGE USE MONTHS PER YEAR	EXPENSE PER WORKING MONTH PERCENT	APPLICATION OF A.G.C. SCHEDULE TO OWNER'S VALUES	
	Depreciation	Overhauling, Major Repairs Painting	Interest Taxes Storage Insurance	Total Ownership Expense			Value Dollars	Expense Per Working Month Dollars
							(Fill in your own values)	

CRANES (Cont'd.)

Truck Crane

Diesel or gasoline, 4—6 tons, 10 ft. radius	25	15	11	51	8	6.4	_____	_____
8—12 tons, 10 ft. radius...	20	15	11	46	8	5.8	_____	_____
15—25 tons, 10 ft. radius .	17	10	11	38	7	5.4	_____	_____
30—50 tons, 10 ft. radius .	13	10	11	34	7	4.9	_____	_____
50—70 tons, 12 ft. radius .	13	10	11	34	7	4.9	_____	_____
80—100 tons, 12 ft. radius	15	11	11	37	7	5.3	_____	_____
110—125 tons, 12 ft. radius	14	10	11	35	7	5	_____	_____

DERRICKS

Including winch, rope, and load block

Circle swing, 1,000 to 3,000 lbs.	13	15	11	39	7	5.6	_____	_____

Guy type:

Capacity   Boom   Mast

3—50 tons, all sizes	8	12	11	31	7	4.4	_____	_____

LOADERS

Belt Loader, dozer fed, 48—60 in.	20	15	11	46	6	7.7	_____	_____

Bucket Type

1—2-1/2 cu. yd. per min.	20	15	11	46	6	7.7	_____	_____
3—8 cu. yd. per min.	20	12	11	43	6	7.2	_____	_____
to 20 cu. yd. per minute..........................	19	11	11	41	7	5.9	_____	_____

Front End, crawler tractor:

For tractor h.p.   Bucket

From   To   Cap. cu. yd.

20 40 1/2—1-1/8	33	15	11	59	8	7.4	_____	_____
60 85 1-1/2—3-1/2	25	15	11	51	8	6.4	_____	_____
85 135 2.7—9	20	12	11	43	6	7.2	_____	_____
135 and up.................................	20	12	11	43	6	7.2	_____	_____

EQUIPMENT	AVERAGE ANNUAL EXPENSE PERCENT OF CAPITAL INVESTMENT WITHOUT FIELD REPAIRS				AVERAGE USE MONTHS PER YEAR	EXPENSE PER WORKING MONTH PERCENT	APPLICATION OF A.G.C. SCHEDULE TO OWNER'S VALUES	
	Depreciation	Overhauling, Major Repairs Painting	Interest Taxes Storage Insurance	Total Ownership Expense			Value Dollars	Expense Per Working Month Dollars
							(Fill in your own values)	

LOADERS (Cont'd.)

Front end, for wheel type tractor:

Complete with accessories

Bucket

Cap. cu. yd.

10—17-1/2 cu. ft.	33	15	11	59	8	7.4	_____	_____
3/4—1-1/8 cu. yd.	33	15	11	59	8	7.4	_____	_____
1-1/4—1-3/4 cu. yd.	25	15	11	51	8	6.4	_____	_____
2—3 cu. yd.	25	15	11	51	8	6.4	_____	_____
3—5 cu. yd.	20	12	11	43	6	7.2	_____	_____
5—7 cu. yd.	20	12	11	43	6	7.2	_____	_____

SCRAPERS

Self-loading Scoop Pan

Pneumatic tires—without power unit

Capacities

Struck	Heaped								
2—3 cu. yds.	3—5 cu. yds.	25	15	11	51	8	6.4	_____	_____
4—36 cu. yds.	6—50 cu. yds.	20	15	11	46	8	5.8	_____	_____

Rotary (Drag), 18—27 cu. ft.	33	15	11	59	7	8.4	_____	_____
44—70 cu. ft.	25	15	11	51	7	7.3	_____	_____

SHOVELS AND BACKHOES

Crawler

Diesel or Gasoline, 3/8 cu. yd.	27	15	11	53	9	5.9	_____	_____
1/2 cu. yd.	25	15	11	51	9	5.7	_____	_____
3/4 cu. yd.	23	15	11	49	9	5.4	_____	_____
1 cu. yd.	21	15	11	47	9	5.2	_____	_____
1-1/4 cu. yd.	20	15	11	46	8.5	5.4	_____	_____
1-1/2 cu. yd.	18	15	11	44	8.5	5.2	_____	_____
1-3/4—2 cu. yd.	17	15	11	43	8	5.4	_____	_____

EQUIPMENT	AVERAGE ANNUAL EXPENSE PERCENT OF CAPITAL INVESTMENT WITHOUT FIELD REPAIRS				AVERAGE USE MONTHS PER YEAR	EXPENSE PER WORKING MONTH PERCENT	APPLICATION OF A.G.C. SCHEDULE TO OWNER'S VALUES	
	Depreciation	Overhauling, Major Repairs Painting	Interest Taxes Storage Insurance	Total Ownership Expense			Value Dollars	Expense Per Working Month Dollars
							(Fill in your own values)	
SHOVELS AND BACKHOES (Cont'd.)								
2-1/4–2-1/2 cu. yd.	16	15	11	42	8	5.3		
3-1/2 cu. yd.	14	15	11	40	7.5	5.3		
4 cu. yd.	14	15	11	40	7.5	5.3		
4–6 cu. yd.	14	15	11	40	6	6.7		
Truck Mounted, diesel or gasoline								
Pull type								
3/8 cu. yd.	25	15	11	51	9	5.7		
1/2 cu. yd.	23	15	11	51	9	5.7		
3/4–1-1/2 cu. yd.	21	15	11	47	9	5.2		
1-1/2 cu. yd. and up	19	15	11	45	9	5.0		
Hydraulic								
1/2 cu. yd.	23	15	11	49	9	5.4		
3/4–1-1/2 cu. yd.	20	19	11	46	9	5.1		
1-1/2 cu. yd. and up	20	15	11	46	8	5.8		
TRACTORS								
Crawler, diesel engine, gear drive								
Drawbar h.p.								
From To								
20 52	25	15	11	51	9	5.7		
53 265	20	15	11	46	8	5.8		
266 500	20	15	11	46	6	7.7		
Crawler, gasoline engine								
Drawbar h.p.								
From To								
20 32	33	15	11	59	9	6.6		
33 41	30	15	11	56	9	6.2		
42 52	28	15	11	54	9	6.0		
53 66	25	15	11	51	9	5.7		
67 105	20	15	11	46	8	5.8		
106 200	20	15	11	46	7	6.6		

EQUIPMENT	AVERAGE ANNUAL EXPENSE PERCENT OF CAPITAL INVESTMENT WITHOUT FIELD REPAIRS				AVERAGE USE MONTHS PER YEAR	EXPENSE PER WORKING MONTH PERCENT	APPLICATION OF A.G.C. SCHEDULE TO OWNER'S VALUES	
	Depreciation	Overhauling, Major Repairs Painting	Interest Taxes Storage Insurance	Total Owner-ship Expense			Value Dollars	Expense Per Working Month Dollars
							(Fill in your own values)	
TRACTORS (Cont'd.)								
Four-Wheeled, rubber tired								
Diesel engine, direct drive								
Brake h.p.								
From To								
35 47	30	15	11	56	9	6.2	_____	_____
48 60	25	15	11	51	9	5.7	_____	_____
93 160	20	15	11	46	8	5.8	_____	_____
Gasoline engine								
Brake h.p.								
From To								
10 16	50	15	11	76	9	8.4	_____	_____
17 24	40	15	11	66	9	7.3	_____	_____
25 33	33	15	11	59	9	6.6	_____	_____
34 45	30	15	11	56	9	6.2	_____	_____
46 60	25	15	11	51	9	5.7	_____	_____
61 75	25	15	11	51	8	6.4	_____	_____
TRUCKS								
Diesel, stake body, 2 ton..........................	25	12	11	48	9	5.3	_____	_____
3-1/2—5 ton..................	20	12	11	43	8	5.4	_____	_____
10 ton	15	12	11	38	8	4.8	_____	_____
dump body, 2 ton..........................	25	15	11	51	9	5.7	_____	_____
3-1/2—5 ton..................	20	15	11	46	8	5.8	_____	_____
Diesel, heavy duty, dump body, 5—12 cu. yd. ...	20	12	11	43	8	5.4	_____	_____
15—18 cu. yd. .	15	12	11	38	8	4.8	_____	_____
18—22 cu. yd. .	17	14	11	42	8	5.3	_____	_____
22—28 cu. yd. .	17	15	11	43	8	5.4	_____	_____
28—35 cu. yd. .	18	15	11	44	8	5.5	_____	_____
35—42 cu. yd. .	17	15	11	43	8	5.4	_____	_____
42—50 cu. yd. .	17	15	11	43	8	5.4	_____	_____
50—62 cu. yd. .	17	15	11	43	8	5.4	_____	_____
62—74 cu. yd. .	17	15	11	43	8	5.4	_____	_____

EQUIPMENT	AVERAGE ANNUAL EXPENSE PERCENT OF CAPITAL INVESTMENT WITHOUT FIELD REPAIRS				AVERAGE USE MONTHS PER YEAR	EXPENSE PER WORKING MONTH PERCENT	APPLICATION OF A.G.C. SCHEDULE TO OWNER'S VALUES	
	Depreciation	Overhauling, Major Repairs Painting	Interest Taxes Storage Insurance	Total Ownership Expense			Value Dollars	Expense Per Working Month Dollars
							(Fill in your own values)	

WELDING MACHINES AND ACETYLENE TORCHES

Acetylene Torch

Complete working unit, all sizes	25	10	11	46	8	5.8	_____	_____

Welding Equipment

Acetylene hand truck	25	20	11	56	8	7.0	_____	_____
Acetylene hand case.............................	25	20	11	56	8	7.0	_____	_____
Welding machines								
A.C. Arc (75–175 Amp.) to (750–1,000 Amp.)	20	12	11	43	8	5.4	_____	_____
D.C. Arc, Gasoline engine driven (150 to 250 Amp.) to (500 to 700 Amp.)	20	15	11	46	8	5.8	_____	_____
D.C. Arc, 60 cycle electric motor (150 to 250 Amp.) to 700 to 900 Amp.).....................	20	12	11	43	8	5.4	_____	_____
D.C. Arc, 25 cycle electric motor (100 to 250 Amp.) to (700 to 900 Amp.).....................	20	12	11	43	8	5.4	_____	_____
D.C. Arc, D.C. electric motor (100 to 250 Amp.) to (500 to 700 Amp.)....................	20	12	11	43	8	5.4	_____	_____
D.C. Arc, Diesel engine driven (200 to 350 Amp.) to (350 to 500 Amp.)....................	20	10	11	41	8	5.1	_____	_____
A.C./D.C. Arc, Transformer Rectifier (175 to 275 Amp.) to (550 to 750 Amp.)...........	17	10	11	38	8	4.8	_____	_____
D.C. Arc, Transformer Rectifier (175 to 275 Amp.) to (450 to 550 Amp.)...................	17	10	11	38	8	4.8	_____	_____

283

Appendix F

QUANTITY TAKEOFF

Estimate by _____   Date _____   Ckd. by _____   Date _____

Item	Identity	Location	Quantity Computations	Total	Units

SUMMARIES & UNIT COSTS

Estimate by _____ Date _____ Ckd. by _____ Date _____

Item No.	Identity & Cost Source	Computation of Unit Costs	Total	Units

DIRECT COSTS

Estimate by _____ Date _____ Ckd. by _____ Date _____

Item No.	Identity & Location	Quantity		Unit Cost Each			Total Cost Each			Total Cost
		No.	Unit	Equip.	Mat'l.	Labor	Equip.	Mat'l.	Labor	

OVERHEAD & PROFIT

Estimate by _____ Date _____ Ckd. by _____ Date _____

Item No.	Class of Expense	Computations of Overhead Expense	Total Cost
1600	Gen. overhead (% direct cost)		
	Job overhead		
1700	Int. on operating capital		
1701	Superintendent's salary		
1702	Supt. pickup truck - rental		
	do. operating cost		
1703	Job trucks - rental		
	do. operating cost		
	do. wages - driver's		
1704	Lifting equipment		
	do. operating cost		
	do. wages - operator's		
1705	Job office - rental		
	do. salaries		
	do. supplies		
1706	Utilities & connections		
1707	Social Security		
1708	Workmen's Compensation		
1709	Pub. Lia. & Prop. Damage		
1710	Fed. & State Unemp. Ins.		
1711	Patents & royalties		
1712	Barricades		
1713	Temporary toilets		
1714	Cut and patch for trades		
1715	Permits		
1716	Protection adjacent prop.		
1717	Final cleanup		
	Subtotal of computed overhead		
1718	Contingencies (% of computed overhead)		
	Total overhead		
	Direct cost from Direct Cost sheet		
	Subtotal (Total overhead plus direct cost)		
	Profit		
	Total cost		
	Performance bond		
	Total amount of bid		

Index

Index